The *Silent* Takeover

The *Silent* Takeover

Global Capitalism and the Death of Democracy

NOREENA HERTZ

HarperBusiness
An Imprint of HarperCollins*Publishers*

HarperCollins books may be purchased for educational, business, or sales promotional use. For information please write: Special Markets Department, HarperCollins Publishers Inc., 10 East 53rd Street, New York, NY 10022.

First HarperBusiness paperback edition published 2003

The Library of Congress has catalogued the hardcover edition as follows:

Library of Congress Cataloging-in-Publication Data
Hertz, Noreena.
The silent takeover : global capitalism and the death of democracy / Noreena Hertz.
p. cm.
Includes bibliographical references and index.
1. Social responsibility of business. 2. Corporations—Social aspects. 3. Globalization. 4. Capitalism. 5. Democracy. I. Title.
HD60 .H47 2002
338.8'8—dc21 2002021452

ISBN 0-7432-3478-2

ISBN 0-06-055973-X (pbk.)

03 04 05 06 07 ❖ / RRD 10 9 8 7 6 5 4 3 2 1

For my late mother—Leah, my father—Jonathan, and sister—Arabel.

With deepest love and gratitude.

CONTENTS

1

The Revolution Will Not Be Televised

Dancing with the Pink Fairies

J-20. For those in the know the acronym is easily decipherable: July 20, 2001, the call for action transmitted to hundreds of thousands at the click of a mouse. J-20—Genoa.

I first learned about the Genoa protests through the Net, as did most of those who gathered there. A chain letter sent to thousands and forwarded to thousands more eventually reached me. Cyberwar with a clear message: Be there if you think that globalization is failing. Be there if you want to protest against global capitalism. If you think multinational corporations are too powerful. If you no longer believe your elected representatives will listen. Be there if you want to be heard.

On July 20 Genoa was host city to the G8 annual summit, and the place to be for the "veterans" of Seattle, Melbourne, and London's City and Parliament Square riots, for the veterans of Washington, Prague, Nice, Quebec, and Gothenburg (if "veterans" is the appropriate term for a movement only a couple of years old). They flocked there in droves: pink fairies in drag, red devils handing out "Boycott Bacardi" leaflets, Italian anarchists in game-show padded body armor, environmentalists with mobile phones, suburbanites with cameras snapping as if they were on a day trip to the big city—a babel of different languages and different objectives gathered under the one "anti" banner, a babel that would continue to make its voice heard in Florence in 2002, and in Porte Alegre, Evian, and London in 2003.

I was prepared for the tear gas: I had read the California-based Ruckus Society's handbook, required reading for protesters, and had brought the requisite lemon and vinegar and a handkerchief to wrap around my face, as well as fake blood in a traveling shampoo container (good when you want to get let through a crowd). I was prepared for the police standoffs: I had studied the tactics of civil disobedience and direct action at the nonviolence workshop I had attended earlier that year in a hangarlike meeting place on the northwest outskirts of Prague. Although nothing could have fully primed me for the brutality of the Italian police.

What I was not prepared for was the extent of the sense of community among the divergent and often conflicting interests, the sense of camaraderie and unity around a shared opposition to the status quo. Neither was I prepared for the sheer rage, inflamed by the insistent drumming and by the mournful wailing of the rainbow-stringed whistles sold at a dollar a piece: the black bloc anarchists intent on smashing shop front windows; the focus of many around me on tearing down the fence that the Italian authorities had erected to keep the world leaders in and the world protestors out.

Least of all, perhaps, was I prepared for the extent to which those I spoke with were utterly disillusioned with politics and politicians, corporations and businesspeople alike, and the lengths to which they were prepared to go to break what they saw as a conspiracy of silence. The bare-chested young man with arms splayed in the sign of a pacifist, who remained upright despite the fire of a water cannon pounding against his back; Venus, the girl with pink hair and glitter stars stuck on her eyes, who told me in a soft Irish lilt that she was "willing to die for this cause."

Twelve years after the tanks last drove onto Red Square, fourteen years after the Berlin Wall came down, after the longest period of economic boom in modern times, dissent is nevertheless growing at a remarkable rate, voiced not only by the hundreds of thousands who gathered in Genoa or Gothenburg, Prague or Seattle, Florence or Porte Alegre, not only by the rainbow warriors, but by disparate and often surprising parties—ordinary people with ordinary lives, homemakers, schoolteachers—suburbanites, city dwellers, and establishment figures, too. George Soros, the legendary financier, now considers himself part of the anti-globalization movement; Joseph Stiglitz, the Nobel Prize–winning former chief economist of the World Bank, is now one of Washington's most vocal critics. And added to these, in the wake of Adelphia, Enron, Tyco, Worldcom, are trade unionists, pensioners, former employees of bankrupt companies, shareholders, and retiree organizations.

All over the world, concerns are being raised about governments' loyalties and corporations' objectives. Concerns that the pendulum of capitalism may have swung just a bit too far; that our love affair with the free market may have obscured harsh truths; that too many are losing out. That the state cannot be trusted to look after our interests; and that we are paying too high a price for our increased economic growth. They are worried that the sound of business is drowning out the voices of the people.

The fairy-tale ending of the story that began in Westminster on May 3, 1979, the day Margaret Thatcher came into power, and was later reproduced in the United States, Latin America, East Asia, India, most of Africa, and the rest of Europe—the story of the streets being paved with gold, and the realization of the American dream—is no longer taken for granted. Myths that were perpetuated during the cold war era, out of fear of weakening "our" position, are beginning to be debunked. Wealth doesn't always trickle down. There are limits to growth. The state will not protect us. A society guided only by the invisible hand of the market is not only imperfect, but also unjust.[1]

The world that is emerging from the cold war is the antithesis of the shrink-wrapped One World of the hyperglobalists. It is in fact confused, contradictory, and mercurial. It is a world in which a litany of doubts is starting to be recited, not at the ballot box, but in cathedrals, shopping malls, and on the streets. A world in which loyalties can no longer be determined, and allegiances seem to have switched. While BP was running a program for its top two hundred executives on the future of capitalism in which the merits and demerits of globalization were debated, a British Labour government was fighting to privatize air traffic control.

The *Space Odyssey* world of 2003 is getting dangerously close to the apocalyptic visions of *Rollerball, Network,* and *Soylent Green.* It is a world in which, as we will see, corporations are taking over from the state, the businessman becoming more powerful than the politician, and commercial interests are paramount. As I will show, protest is fast becoming the only way of affecting the policies and controlling the excesses of corporate activity.

The Benetton Bubble

We can date the beginning of this world, this world of the Silent Takeover, from Margaret Thatcher's ascendency. The hairspray-helmeted Iron Lady proselytized a particular brand of capitalism with her *compadre* Ronald Reagan that put inordinate power into the hands of corporations, and gained market share at the expense not only of politics but also of democracy. And it has been a durable product. Apart from a few discreet tweaks, theirs remains the dominant ideology across much of the world. Politics in the post–cold war age has become increasingly homogenized, standardized, a commodity.

Benetton provides an apt metaphor for politics today. Over the last nineteen years this Italian fashion company has run the most provocative advertising campaigns ever seen. Twenty-foot billboards with the picture of a starving black baby; the AIDS victim at his moment of death; the bloodied uniform of a dead Bosnian soldier; the "United Killers of Benetton" campaign, a ninety-six-page magazine insert with photograph after photograph of condemned prisoners languishing on America's death rows. Benetton shocked us to attention, but shock is all it provided. It didn't rally us into action. Nor did it try and address these issues itself. Their advertising provided no exploration of the morality of war, there was no attempt to relieve poverty or cure AIDS. The only goal was to increase sales, not to start a discussion of the issues behind capital punishment. And if it profited from others' misery, so what?[2]

We are living in a Benetton bubble. We are presented with shocking images by politicians who try to win our favor by demonizing their opponents and highlighting the dangers of the "wrong" representation. They speak of making a difference and changing our lives. Mainstream parties offer us supposedly different solutions and choices: Democrats tout liberal virtues, Republicans tout conservatism, all in an attempt to secure our votes.

But the rhetoric is not matched by reality. The solutions our politicians offer are as bogus as those of Benetton: a Chinese girl standing next to an American boy, a black woman holding hands with a white woman. Models with unusual faces, strong faces, sometimes beautiful, sometimes not. Multicolored people in multicolored clothes.

Political answers have become as illusory as the rows and rows of homogenized clothes, standard T-shirts, and cardigans folded in your local Benetton store. Commercialized conservatism and conformity par excellence. Politicians offer only one solution: a system based on laissez-faire economics, the culture of consumerism, the power of finance and free trade. They try and sell it in varying shades of blue, red, or yellow, but it is still a system in which the corporation is king, the state its subject, its citizens consumers. A silent nullification of the social contract.

But, I will argue, the system is undeniably failing. Behind the ideological consensus and supposed triumph of capitalism, cracks are appearing. If everything is so wonderful, why, as we will see, are people ignoring the ballot box and taking to the streets and shopping malls instead? How meaningful is democracy if only half the people turn out to vote, as in the Bush–Gore presidential election, even though everyone knew it was going to be a close race? What is the worth of representation if, as I will show, our politicians now jump to the commands of corporations rather than those of their own citizens?

Capitalism on Tap

It took time for people to rise up in protest, to see that the weightless state was unlikely to deliver the clean, safe world that they wanted their children to grow up in. For a long time people didn't question the one-ideology, homogeneous world. Why should they? For many, life was good and getting better. For most of the past twenty years the stock market rose and interest rates fell. More people than ever before own their own homes. Two thirds of us, in the developed world, have television sets of our own.[3] Most of us, in the West that is, have cars. Our children wear Nike and Baby Gap. The middle class has grown and grown.

We are drip-fed images that reinforce this capitalist dream. Studios and networks beatify the very essence of capitalism. Prevailing norms and mainstream thoughts are recorded, replayed, and reinforced in

Technicolor, while any criticism of the orthodoxy is consciously quashed. The peaceful element in the protests of Seattle, Gothenburg, and Genoa hardly made it to our screens. Proctor & Gamble explicitly prohibits programming around its commercials "which could in any way further the concept of business as cold or ruthless."[4] Programs are sought that reinforce the advertisers' message. "Each time a television set is turned on, the political, economic, and moral basis for a profit-driven social order is implicitly legitimised."[5]

In 1997 Adbusters,[6] a Canadian "culture-jamming" organization, tried to air a counter-consumerism ad in which an animated pig superimposed on a map of North America smacked its lips while saying, "The average North American consumes five times more than a Mexican, ten times more than a Chinese person and thirty times more than a person from India. . . . Give it a rest. November 28 is Buy Nothing Day." But U.S. stations such as NBC, CBS, and ABC flatly refused to run it, even though the funding for it was there. "We don't want to take any advertising that's inimical to our legitimate business interests," said Richard Gitter, vice president of advertising standards at General Electric Company–owned NBC.

Westinghouse Electric Corporation's CBS went even further in a letter rejecting the commercial, justifying its decision on the grounds that Buy Nothing Day was "in opposition to the current economic policy in the United States."[7]

Corporate Behemoths

Such is our legacy. A world in which consumerism is equated with economic policy, where corporate interests reign, where corporations spew their jargon on to the airwaves and stifle nations with their imperial rule. Corporations have become behemoths, huge global giants that wield immense political power.

Propelled by government policies of privatization, deregulation, and trade liberalization, and the technological developments of the past twenty years, a power shift has taken place. The hundred largest multinational corporations now control about 20 percent of global foreign assets, and fifty-one of the one hundred biggest economies in the world are now corporations (twenty-nine out of the top hundred, if measured in value-added terms).[8] The sales of General Motors and Ford are greater than the GDP of the whole of sub-Saharan Africa; the assets of IBM, BP, and General Electric outstrip the economic capabilities of most small nations; Wal-Mart, the supermarket retailer, has higher revenues than most Central and Eastern Europe states; and Exxon is comparable in economic size to the economies of Chile and Pakistan.[9]

The size of corporations is increasing. In the first year of the new millennium, Vodafone merged with Mannesmann (a purchase worth $183 billion), Chrysler with Daimler (the merged company now employs over 400,000 people), Smith Kline Beecham with Glaxo Wellcome (now reporting pretax profits of $7.6 billion as Glaxo-SmithKline), and AOL with Time Warner in a merger worth, at the time, $350 billion—five thousand mergers in total in 2000 and double the level of a decade earlier. These megamergers mock the M&A activity of the 1980s. Each new merger is bigger than the one before, and governments rarely stand in the way. Each new merger gives corporations even more power. All the goods we buy or use—our gasoline, the drugs our doctors prescribe, essentials like water, transport, health, and education, even the new school computers and the crops growing in the fields around our communities—are in the grip of corporations which may, at their whim, nurture, support, or strangle us.

This is the world of the Silent Takeover, the world at the dawn of the new millennium. Governments' hands appear tied and we are increasingly dependent on corporations. Business is in the driver's seat, corporations determine the rules of the game, and governments have become referees, enforcing rules laid down by others. Portable corporations are now movable feasts and governments go to great lengths to attract or retain them on their shores. Blind eyes are turned to tax loopholes. Business moguls use sophisticated tax dodges to keep their bounty offshore. Rupert Murdoch's News Corporation pays only 6 percent tax worldwide; and in the U.K., up to the end of 1998, it paid no net British corporation tax at all, despite having made £1.4 billion profit there since June 1987.[10] This is a world in which, although we already see the signs of the eroding tax base in our crumbling public services and infrastructure, our elected representatives kowtow to business, afraid not to dance to the piper's tune.

Governments once battled for physical territory; today they fight in the main for market share. One of their primary jobs has become that of ensuring an environment in which business can prosper, and which is attractive to business. The role of nation states has become to a large extent simply that of providing the public goods and infrastructure that business needs at the lowest costs while protecting the world's free trade system.

Divided We Fall

In the process, justice, equity, rights, the environment, and even issues of national security fall by the wayside. Take the case of the Taliban—supported by the United States until 1997 because of U.S. oil company interests, despite the regime's dismal human rights record. Social justice has come to mean access to markets. Social safety nets have been weakened. Union power has been smashed.

Never before in modern times has the gap between the haves and the have-nots been so wide, never have so many been excluded or so championless. Forty-five million Americans have no health insurance. In Manhattan people fish empty drink cans and bottles from trash cans to claim their five cents' redemption value, while in London, car windshield washers armed with squeegees and pails of dirty water ambush drivers at traffic lights. Americans spend $8 billion a year on cosmetics while the world cannot find the $9 billion the UN reckons is needed to give all people access to clean drinking water and sanitation. The British Labour party has gone on record as saying that wealth creation is now more important than wealth redistribution.[11]

In America, during the ten years after 1988, income for the poorest families rose less than 1 percent, while it jumped 15 percent for the richest fifth. In New York City the poorest 20 percent earn an annual average of $10,700 while the wealthiest 20 percent earn $152,350.[12] Wages for those at the bottom are so low that, despite the country's low unemployment figures, millions of employed Americans and one in five American children are now living in poverty, with 48 percent of America's hungry being families with children. Never since the 1920s has the gap between rich and poor been so great.[13] Bill Gates's net worth alone at the end of the last century, for example, equaled the total net worth of the bottom 50 percent of American families.[14]

Capitalism has triumphed, but its spoils are not shared by all. Its failings are ignored by governments which, thanks to the very policy measures they introduced, are increasingly unable to deal with the consequences of their system.

And that system is rotten. Political scandals are unveiled all too frequently: Kohl, Schmidt, and Mitterrand are among those we already know or suspect. Even those politicians not on the take are increasingly indebted to or enmeshed with business.

Nowhere is this more apparent than in the United States. Clinton's presidency was immersed in scandal at once: from the Whitewater allegations, via overnight stays in the Lincoln bedroom for party funders, to the final act of pardoning tax evader and arms dealer Marc Rich. For candidates for the 2000 American presidential election, their very ability to run depended upon their securing corporate funding. Al Gore's campaign war chest was $133 million; George W. Bush's $191 million,[15] with Enron and its executives among Bush's biggest donors, and many of his closest advisers—Cheney, Rove, Daniels, and others—have clear and obvious links to big business. And objections to the McCain–Feingold bill on campaign finance reform, which, once in effect, would ban businesses, trade unions, and individuals from making unlimited "soft money" contributions to American political parties, came from both Democrats and Republicans.

No wonder the politician's star is fading. People recognize politicians' conflicting interests and unwillingness to champion them, and are beginning to abandon politics en masse. Whereas the 1980s saw democracy emerging all over the world as the dominant mode of government, imbued with a unique legitimacy and commanding mass support, by the 1990s voter turnout almost everywhere was falling, party membership declining, and politicians rated below meter maids as worthy of respect.[16] All over the world, from the old democracies of the United States and Western Europe to the young nations of Latin America and the Far East, people have less confidence in the institutions of government today than they had a decade ago. Only 59 percent of British voters voted at the 2001 general election, down from 69 percent in 1997, the lowest turnout since World War I. In the United States, not in nearly two centuries have so many American citizens freely abstained from voting as in the past seven years.[17] The product sold by politicians is seen as broken, no longer deemed worth buying.

Breaking the Silence

This is the world of the Silent Takeover that I will explore in this book. My aim is to make sense of this world and understand where it is likely to take us: a world in which corporate resources dwarf those of nations, and businessmen outrank politicians; in which three quarters of Americans now think that business has gained too much power over many aspects of their lives;[18] and in which, despite the ever harder sell of party politics, fewer and fewer trouble to vote. Economics is now accorded greater respect than politics, the citizen has been abandoned, and the consumer is all that matters. "Participation in the market has [been] substituted for participation in politics."[19]

My argument is not intended to be anticapitalist. Capitalism is clearly the best system for generating wealth, and free trade and open capital markets have brought unprecedented economic growth to most if not all of the world. Nor is the book intended to be antibusiness. Corporations are not amoral but, I will argue, they are morally ambivalent. In fact, under certain market conditions, business is more able and willing than government to take on many of the world's problems. "Social responsibility," "sustainable development," and "environmental impact" are terms more likely to be heard today from CEOs than from government ministers.

Neither do I intend to glorify government. Although, as I will argue, the state has a clear role to play in society, I remain highly skeptical of government's ability to play this role, especially now that the boundaries between business and government have blurred so much and there is such a lack of true political leadership or will.

What my book *is* intended to be, however, is unashamedly propeople, prodemocracy, and projustice. I mean to question the moral justification for a brand of capitalism that encourages governments to sell their citizens for a song; to challenge the legitimacy of a world in which many lose and few win; to reveal how the takeover endangers democracy; and to argue that there is a fundamental paradox at the heart of laissez-faire capitalism, that by reducing the state to its bare minimum and putting corporations at center stage the state risks jeopardizing its own legitimacy. I will explore the implications of a world in which we cannot trust governments to look after our interests, in which even the motives for going to war with Iraq were clouded by big business interests, and in which unelected powers—big corporations—are taking over governments' roles; and I will examine the consequences of a political mind-set which values the pursuit of market share above all else. I will chart the unfettered pursuit of profit, and confront those who justify pork-barrel politics as an expression of free speech, and those who justify nonintervention in other countries' affairs for reasons of their own trade interests.

Over the last two decades the balance of power between politics and commerce has shifted radically, leaving politicians increasingly subordinate to the colossal economic power of big business. Unleashed by the Reagan–Thatcher axis, and accelerated by the end of the cold war, this process has grown hydralike over the last two decades and now manifests itself in what are diverse positive and negative forms. Whichever way we look at it, corporations are taking on the responsibilities of government.

And as business has extended its role, it has, as we shall see, actually come to define the public realm. The political state has become the corporate state. Governments, by not even acknowledging the take-over, risk shattering the implicit contract between state and citizen that lies at the heart of a democratic society, making the rejection of the ballot box and the embracing of nontraditional forms of political expression increasingly attractive alternatives. Exploring these developments and their consequences will make up the body of this book.

My decision to write *The Silent Takeover* was not a disinterested one. I needed to make sense of my own growing discontent, my own feelings that things were going awry. How could it be that life had in many ways never been better, yet I and so many around me seemed so troubled? How was it that I, the daughter of a woman who devoted much of her life to putting women into politics, now saw politics as a coopted, increasingly meaningless arena—a sideshow whose best act was the farce of the last U.S. presidential elections? How could it be that ten years after landing in Leningrad to set up

Russia's first stock exchange—a traveling saleswoman with an M.B.A. from Wharton in my briefcase—I now felt a burning need to question its very tenets? Why was it that at Cambridge University's business school, where I teach, when I made it clear that I was willing to supervise on the issues that this book examines, I was deluged with so many requests from students that I couldn't possibly satisfy them all?

We stand today at a critical juncture. If we do nothing, if we do not challenge the Silent Takeover, do not question our belief system, do not admit our own culpability in the creation of this "new world order," then all is lost. As we shall see, inequality of income is bad not only for the poor, but for the rich, too.[20] The steady erosion of government and politics is dangerous for all, regardless of political persuasion. A world in which George W passes law after law favoring the interests of big business, Rupert Murdoch has more power than Tony Blair, and corporations set the political agenda is frightening and undemocratic. The idea of corporations taking over the roles of government might in some ways seem appealing, but risks leaving us increasingly without recourse.

The story will be told through a cast of characters that we shall meet on the way. Granny D, the ninety-one-year-old grandmother who walked across America to champion campaign finance reform; Sister Patricia Marshall, the shareholder activist nun who persuaded PepsiCo to sell off its Burmese bottling plant; Oskar Lafontaine, the former German finance minister whose parting comment on his resignation was, "The heart is not traded on the stock market yet." These are but a few of the voices we will hear.

But this book is not just the sum of their disparate stories; it is the sum of all our stories. We are all in the midst of a corporate takeover, and no gated communities or six-figure salaries will protect us from its impact.

My subject is how the Silent Takeover crept up on us, why it matters, and what we can do about it.

2

Living in a Material World

Boogie Woogie in Bhutan

The kingdom of Bhutan, mythical Land of the Thunder Dragon, last of the independent Himalayan principalities, lies between Tibet and India. Wilfully isolationist, it has managed for centuries to follow its own path. Its population of around 600,000 is among the poorest in the world in terms of GNP per capita—average annual income is $550—but this is a misleading picture, as it disregards the fact that more than 85 percent of the population is involved in subsistence farming, and barter transactions are the norm. People in Bhutan are well fed and clothed, and homelessness is virtually non-existent.

"Success" here is determined on the basis of ecological, ethical, and spiritual development; morality and enlightenment are valued above material wealth. Buddhist values are upheld and traditions maintained: the country's two thousand monasteries are active, and Driglam Namsha, the ancient code of conduct, remains part of the school curriculum. According to Bhutan's king, Jigme Singye Wangchuk, "Gross National Happiness is more important than Gross National Product."

The path of development has been carefully managed so as to remain consistent with the country's integral belief system. Unlike its neighbor Nepal, which admitted 500,000 tourists in 1998, Bhutan took in only 6,000 that year; and each of them was provided with a strict code of conduct that included the prohibition of tipping and distributing sweets or pens to local children, to discourage begging.

Another significant generator of foreign revenue, commercial logging, has also been shunned because of the damage it would do to the environment—Bhutanese Buddhism lays great importance on ecology. As C. Dorji, the minister of planning, puts it, "We will not be rushed into an uncritical adoption of all things that are modern; we will draw on the experience of those who have trod the path of development before us, and undertake modernization with caution at a pace consistent with our capacity and needs. We therefore seek to preserve our culture, traditions, value systems, and institutions."

But the tentacles of global capitalism are far-reaching, and they reach even Bhutan, which cannot escape the one-kilowatt broadcast signals that now bounce between the thousands of satellite dishes that have over the past few years been appearing between the prayer flags and prayer wheels that dot the landscape.

Already the impact of the West is apparent. Basketball has replaced archery as the national sport, thanks to the videotapes of NBA games that the king has shipped to him from New York. *Boogie Woogie,* a game show sponsored by Colgate, now rivals the panoramic Himalayan vista for viewers' attention. *Friends, Teletubbies,* BBC, and CNN entertain, inform, and brief. Nightclubs intercut N'Sync and Britney Spears with 1980s Wham and Culture Club. A modern telecommunications system has been put in place and e-mail is replacing letter writing, despite the ten days of free mail service that Queen Tashi Dorji Wangmo Wangchuk, the eldest of the king's four wives, offered the Bhutanese to combat this very development. Children now make pilgrimages to monasteries offering prayers and lighting butter lamps while clad in Spice Girls T-shirts. Farmers sell apples, oranges, potatoes, and cardamon to their Indian and Bangladeshi neighbors for foreign currency; and there are twenty-five privately owned video stores in the capital city, Thimphu.

So even Bhutan, the last Shangri-la, is being infiltrated. Unable to resist the spoils of the West, unable to continue its isolationist policies, it admits Western influences. The situation raises many questions. How soon before the forces of globalization and free market capitalism are irreversibly entrenched? How blindly will the population follow their unelected leader once their eyes are opened to the multichannel, multiparty universe? Will Bhutan really be able to travel "the Middle Path," a way forward that claims to be able to embrace modernity without compromising traditional ideology, or will the worship of Mammon and MTV culture fast replace the homage paid to the Buddha? Are the days of enlightenment and disregard of material success numbered, now that the Bhutanese will be able to see just how little they can afford relative to others? For how much longer will the king and his queens be able to continue to direct the economy and take responsibility for the welfare of their citizens' lives?

If events elsewhere are anything to go by, probably not for very long.

The State in Control

But for now at least, the Bhutanese state remains the principal economic force (most of industry is state-owned) and also the main carer for the people's welfare needs—roles the state played in America and

Europe, too, for much of the past century, before the fundamental mind-set shift of the 1970s and 1980s, when the extreme version of laissez-faire capitalism epitomized by the American model became so dominant, before government fell in love with the free market.

By the end of the nineteenth century, governments in the United States and in Europe had begun to accept that they had responsibilities beyond those of internal order and external security. There was a growing realization that capitalism was responsible for great cruelties, and a sense that the state should play a role in alleviating the harshest elements of the system through some sort of social intervention. And this nascent feeling gained pace over the first half of the twentieth century, through the Wall Street crash and the Great Depression, then World War II, events which brought first mass unemployment, and then even wider human suffering.

In an attempt to address the needs of the poor, and at the same time stave off the threat of communism[1]—the Soviet Union was by now offering its citizens the most generous of welfare packages—by the middle of the last century, most developed states had begun to establish systems of social security and welfare themselves. The components of the package differed between countries, varying from generous schemes for the redistribution of wealth to minimum provision against destitution. But most Western states offered subsidized access to education, health, housing, and personal care services, alongside some form of income maintenance. The dominant mind-set was that no citizen was to be allowed to fall below a minimum standard of overall well-being. In the U.K., for example, between 1945 and the mid-1970s, the proportion of GDP spent on the main welfare services rose from just 5 percent to around 20 percent. Expenditure on the National Health Service rose from about £500 million in 1951 to £5,596 million by 1975.[2] In the United States the expansion of social spending came later. It was not until the Kennedy and Johnson administrations (1961–1969) that the state appeared truly willing to substantially enlarge its welfare provision. And only in 1964, in the midst of considerable prosperity and sustained economic growth, did President Johnson declare a "national war on poverty."[3]

Not only did the state become the main provider of welfare during the postwar period, it also became the key economic actor. In Europe, even before World War II, countries had begun to nationalize industry and this process accelerated after 1945. The electorate was now open to the thought of government controlling the "commanding heights," for they had seen the effectiveness of state control of the wartime economy.

But owning industry was not enough. Postwar governments also felt it legitimate to play an active role in controlling the macroeconomy and the market. The Bretton Woods agreement, signed by the leading industrialized nations in 1944, brought heavy regulation of financial markets. And neoclassical liberalism—a system that had largely left the market alone to regulate economic life—was displaced, in the West at least, by Keynesian economics.

John Maynard Keynes believed that governments could and should intervene in the economy, and evolved a wholly new model that approached the economy from the direction of money and finance. He argued that the economy has no natural tendency to create full employment, so if self-regulation cannot deliver jobs, governments must intervene to provide them. Since depressions result not from spending too much but from spending too little, government spending must be key. His doctrines, conceived in the aftermath of the Wall Street crash and the Great Depression, called for governments to sustain aggregate economic demand and full employment, within the context of a mixed economy and a welfare state. They fitted the moment. The potential for big government to be a force for good had been demonstrated during the war. The social cohesion implied by full employment and an extensive welfare system matched the prevailing mood for stability and security in the hard-won peace.

By the end of the 1940s the British Labour government had fully embraced Keynesian economics. In the United States the 1946 Employment Act committed government to the goal of full employment, although it was not until the 1960s that Kennedy and Johnson moved to an explicitly Keynesian program. Other Western countries, rebuilding their economies after the war, and heavily dependent on American aid, soon adopted a similar model: a large welfare state, state ownership of major industries, and interventionist government. Much of the developing world also embraced state-dominated development strategies.

Rethinking the State

Things began to change at the end of 1973 when the world's major Arab oil-producing states formed a cartel, OPEC, and sent oil prices skyrocketing. With surging oil prices came an upward spiral of prices and wages, triggering economic recession, unemployment, and price inflation of over 20 percent in several countries, and the widespread inability of third world countries to service their debt.

The prevailing doctrine of Keynesianism, which had been so successful in the preceding thirty years, proved unable to cope in these times of trouble. Not only could it not offer any remedy, many believed that it had caused the crisis in the first place. In any case, events made nonsense of one of Keynesianism's most basic tenets: that inflation could not rise at the same time as unemployment. So a new solution was called for as governments came to believe that "the problem lay not in the inefficient management of the prevailing consensus, but in the consensus itself."[4]

The new economic conditions triggered by the oil crisis

demanded a new style of management of the economy: fiscal restraint and control of the money supply. Overnight, almost all Western countries' finance ministers could be heard talking about the need to fight inflation and to rein back the public sector. The providers of loans to countries in crisis made the embracing of this new ethos a condition of providing funds. In Britain, when Labour Chancellor Denis Healey was forced to turn to the International Monetary Fund for a loan in 1976, the reduction of public spending and tight control over inflation were conditions of the IMF's aid.

From that moment Keynesianism, and with it big government, was dying if not already dead. British Prime Minister James Callaghan delivered its epitaph in a speech at the Labour Party conference later that same year: "We used to think you could spend your way out of recession and increase employment by . . . boosting government spending. I tell you in all candor that that option no longer exists."[5] In the United States, President Carter was reaching the same conclusion, cutting public spending in an attempt to stimulate the economy.

So by the late 1970s Keynes, a man whose teachings had been adopted wholesale by the West in an attempt to rebuild a world shattered by war and establish a secure capitalist bloc as a bulwark against communism, was relegated to a footnote in history. Yet despite the abandonment of Keynesianism, it took a few more years for a new form of capitalism with a distinct ideology to triumph. During the Carter and Callaghan administrations, the idea still prevailed that the state existed to resolve contradictions within the market, and was a force for good in the economy.

The Rise of the New Right

The watershed came in 1979 and 1980 with the election first of Margaret Thatcher and then of Ronald Reagan—politicians from the New Right, who enthusiastically advocated the free market and were determinedly hostile to the concept of an interventionist state. Rejecting Keynesianism, the grocer's daughter and the Hollywood actor embraced the views of economists such as Milton Friedman and Friedrich Hayek. These economists didn't dispute that markets could and did fail; but they believed that the free market was capable of allocating goods and services more effectively than the state could, and that government attempts to combat market failures did more harm than good. They harked back to the ideas that had shaped economic policy from the Victorian era through to the Wall Street crash, that "the role of the state was to enforce contracts, to supply sound money . . . to ensure that market forces were not distorted,"[6,7] and, essentially, to provide the best environment for business to flourish, evoking memories of President Calvin Coolidge's dictum, "The business of America is business."

The extent to which this new religion embodied a coherent ideology, a creed of Reaganism or Thatcherism which could be adopted by other states, remains a matter of dispute. The two leaders' goals and priorities were often different. The Reagan administration was dominated by supply-side economists, who advocated tax cuts to give the greatest incentive to production—whereas Thatcher adopted monetarism, emphasizing tight control of the money supply. But there were themes running through the policies of Reagan and Thatcher that gave a discernible character to their politics, and made it possible to identify their followers in other countries. Their views are easiest to define in the negative: as a rejection of all the pillars of the postwar Keynesian consensus. In place of the goals of full employment and a generous welfare state, the New Right favored the reduction of inflation and cuts in public spending[8] (which they regarded as a major cause of the current economic malaise); rather than a mixed economy, they wanted the state cut back to its core, with many of its functions privatized or contracted out.

The New Right felt that too much had been expected from government in the postwar period. Its view was that the role of government should be to alleviate the worst evils of the human lot and provide a framework within which people and communities could pursue their various goals—not, as in previous decades, to positively guarantee general welfare.[9] John Moore, Thatcher's social services secretary, explained in 1987: "For more than a quarter of a century after the last war, public opinion in Britain, encouraged by politicians, traveled down the aberrant path towards even more dependence on an even more powerful state. Under the guise of compassion people were encouraged to see themselves as 'victims of circumstance.' "[10] According to the New Right, the welfare mentality had bred indolence and dependency.

Under these new leaders there was a clear shift of priorities. Interdependence was replaced by independence and egalitarianism was rejected on ideological grounds: the state was no longer to have a role to play in redistributing wealth.[11] Relative standards of poverty were deemed irrelevant; poverty was to be defined by absolute standards of need. As Thatcher argued in 1985, "You are not doing anything against the poor by seeing that the top people are paid well."[12] And the state no longer accepted responsibility to provide unquestioning support for those who for whatever reason were denied an ability to be productive. In 1981, in the aftermath of the first serious riots on the British mainland in the twentieth century, in the London district of Brixton, Secretary of State for Employment Norman Tebbit made the infamous assertion that "My father didn't riot but got on his bike to look for work."[13] "Get on your bike" became the moral imperative of Thatcherism.[14]

And greed was declared good. Oliver Stone's *Wall Street,* Tom Wolfe's *The Bonfire of the Vanities,* Martin Amis's *Money,* and Michael Lewis's *Liar's Poker* faithfully chronicled the times. Power dressing and padded shoulders clad those aspiring to partake in the capitalist dream. Economists of the Chicago school wrote of man as a selfish utility maximizer and, almost in a self-fulfilling prophecy, *Homo economicus,* or economic man, was born.

In the United States the long period of increased involvement of the national government in domestic affairs that began with Roosevelt's New Deal was now over, succeeded by the "New Federalism."[15] Reaganomics (and Thatcherism, too) rested on a firm belief in the "trickle-down" theory, which claims that if the rich are provided with incentives such as lower taxation, they will in turn have more incentive to act as entrepreneurs and so will boost growth and create jobs. Or that if public service industries are turned over to the private sector, they will be run more efficiently and provide more jobs for people who will then start disappearing off the welfare rolls.[16] Providing incentives for the poor to work, such as making welfare less attractive, was also believed to boost economic growth. Eligibility requirements for benefits were tightened, and rights to food stamps and funds from AFDC (Aid to Families with Dependent Children) were withdrawn from some recipients.[17] Unlike in Europe, public ownership had never taken off in the United States, so Reagan's main tool of liberalization was deregulation of the economy, a process kicked off by Jimmy Carter in the 1970s.[18] The Reagan administration cancelled oil price controls, loosened restrictions over railroad transportation, broadcasting, and the oil and natural gas industries, and was reluctant to enforce antitrust legislation.[19] Despite the fact that U.S. trade union leaders had not wielded significant political clout for some time, Reagan echoed Thatcher in a strong commitment to curbing union power. Shortly after assuming office, he was confronted with a strike by the nation's air traffic controllers. He promptly fired them all, substituting military controllers and newly trained workers in their place.[20]

As well as making life considerably easier for the private sector (President Reagan's tax cut bill of 1981 introduced a plethora of new corporate loopholes and set off a wave of corporate tax sheltering, with the result that many large American corporations paid nothing in corporate taxes),[21] Reagan promised to "get government off the backs of the people."[22] Through tax cuts, he aimed to re-create the structure of incentives and rewards that had been frozen by the high-tax policies of his predecessors. The top marginal rate of income tax in the United States fell from 70 percent to 28 percent.[23]

In Britain the new Conservative government of Margaret Thatcher abandoned the ambitions of both the Labour and Conservative governments of the 1950s and '60s, jettisoned the government's commitment to sustain full employment, celebrated the virtues of private rather than public provision, and set itself to reduce the burden of social expenditure, which, it argued, had seriously eroded those economic incentives that alone made sustained economic growth possible.[24,25] The private sector was to be set free, and the state was to be rolled back.

In the U.K. the "family silver"[26] was sold off as Thatcher came to see privatization as the main cure for the ills of the British economy, as well as a convenient way of balancing the budget. A massive sale of assets from the public to the private sector was conducted during the eighties and nineties with the Conservative government raising £67 billion[27] between 1979 and 1997.[28] "In 1979 government institutions owned much or all of coal, steel, gas, electricity, water, railways, airlines, telecommunications, nuclear power, and shipbuilding, and had a significant stake in oil, banking, shipping, and road haulage. By 1997, nearly all of this was in private hands."[29] Secretary of State for Energy Nigel Lawson summed up the party position at that time with his argument in 1982 that "no industry should remain under state ownership unless there is a positive and overwhelming case for it so doing."[30]

Steps were taken to create an economic culture that rewarded enterprise and innovation. Rates of corporate and individual taxation were reduced; price, dividend, and foreign exchange controls were removed with no thought for the vulnerable state in which this would leave the nation. At the Bank of England thousands of people lost their jobs. Restrictions on bank lending and hire purchase were abolished. Controls over broadcasting, telecommunications, transport, and advertising were withdrawn. Right-to-buy schemes for council houses were set up, and shares, not least in the formerly publicly owned utilities, became available much more widely than before. In 1979 there were four times as many trade unionists as shareholders. Within a decade the latter exceeded the former.[31] Capitalism was made "popular"—everyone was to share Thatcher's economic success.[32] Anything or anyone that potentially stood in the way of this success came under attack. Regulation was dismantled, for it was seen as a stranglehold on corporations. The unions were attacked with ferocity and held largely to blame for the poor economic performance of British industry. The denunciation of trade unionism became an article of faith on the New Right.

By the early 1980s the role of government in America and England had fundamentally and irreversibly changed. Free enterprise was seen as the key to economic success, and the task of government was now "to create a framework in which individuals and groups can successfully pursue their respective ends."[33] David Stockman, Reagan's budget director, said, "The . . . vision of a good society rested on the strength and productive potential of free men in free markets."[34] Successful and unfettered corporations would, it was believed, build the road to Nirvana.

Exporting Capitalism

This creed of free market capitalism, Anglo-American style, was soon disseminated across the world. Aided by developments in communications and the media, which ensured that ideas spread quickly, and by the single-mindedness of the neo-liberal international lending institutions, the IMF and the World Bank, who were promoting the so-called Washington consensus, capitalism's foot soldiers marched from Latin America to East Asia, India, and most of Africa, from old and declining capitalist nations such as the U.K. to vigorous capitalist economies with strong traditions of regulation, such as Germany; and eventually even to the former command economies of the Communist world.[35] "The market" became the catchphrase of the 1980s and 1990s as liberalized states bore witness to the benefits of the capitalist system.

The first countries to embrace Anglo-American free market capitalism were Britain's former dominions. Sliding economic performance in Australia in the 1980s led Treasury Minister (later prime minister) Paul Keating to warn that the country risked becoming a "banana republic" if it did not reform. His methods—deregulation, fiscal rectitude, and privatization—were highly reminiscent of Thatcher's. In Canada during the same period, Brian Mulroney liberalized the laws restricting foreign investment in the country, opening up the Canadian market for free trade. In New Zealand, "One of the world's most comprehensive social democracies became a neoliberal state. . . . Uncompromising neo-liberal ideology animated a program of radical reform in which no major social institution was left unreconstructed."[36]

In Latin America the military dictators who dominated politics in the 1980s also showed themselves to be keen disciples of the New Right. In Chile, under General Pinochet, the lack of democratic constraints facilitated the imposition of painful monetarist economic policies carried out under the guidance of a team of economists from the University of Chicago. And by the early 1990s all major Latin American leaders, "President Carlos Salinas de Gortari in Mexico, President Carlos Menem in Argentina, and President Fernando Collor de Mello in Brazil [sought] to implement far-reaching programs of economic liberalization, accepting the need for market competition and openness to the world economy,"[37] believing that their underdevelopment was due to the "insufficient degree of capitalism that had been practised in their countries in the past," and recognizing that their only chance of securing IMF loans was by implementing reform packages along the Washington consensus lines.[38]

Across Europe high levels of inflation and public indebtedness forced governments to question the basis of their economic policies.[39] Helmut Kohl in Germany and Jacques Chirac in France were not New Right zealots like Thatcher and Reagan, yet they appreciated the financial benefits of privatization; and both showed an understanding of the realities of the new global environment in which they, too, would have to display willingness to bring down business taxes and deregulate the labor market, or else lose out on inward investment. So under Kohl "social benefits and health provisions were restricted, companies privatized, strike laws changed in bosses' favor, business taxes trimmed."[40] In Italy and France companies worth roughly $50 billion in all were privatized in the ten years to 1995,[41] resulting in a massive increase in privately held big business.

The End of the Cold War

In the meantime communism, the only other major ideological contender, was dying an ignoble death. In the autumn of 1988 Mikhail Gorbachev traveled to New York to deliver a historic address to the UN General Assembly. The cold war was over, he proclaimed. Communism had failed in its seventy-year battle against the global capitalist system. A year later the Berlin Wall came crashing down. Three years after that, the Soviet Union collapsed.

But while William the Conqueror took Britain with the sword, the Soviet bloc was vanquished by the Coca-Cola bottle. Free market capitalism demolished communism by transmitting Rupert Murdoch and Ted Turner's weltanschauung, making it impossible for

Communist governments to continue to shield their populations from awareness of the prosperity of Western states. McDonald's, Levi's, BMWs, and rock music had become symbols of the Western way of life as important to the East Europeans as multiparty democracy or freedom of speech and travel. And the Soviet government could no longer resist an international capitalist system that had grown more and more wealthy over the preceding two decades. Huge expenditure on the military had taken its toll. Matching Reagan's proposed Star Wars program was a financial impossibility. The Soviet Union's need to be ready to fight the rest of the world had become increasingly untenable.

The collapse of Soviet-style communism in turn affected those states outside Europe that had looked to the Russian model in structuring their economies. In India, for example, which traded extensively with Communist countries, the revolutions in Eastern Europe spurred the liberalization of its own economy.[42] It is now moving ahead with privatization and easing the process of foreign direct investment. In Africa after 1989 Zambia and Tanzania were among several countries that began to convert to more market-minded philosophies.[43] Communism's terminal crisis began in China when the leadership recognized that the country was being left behind by the rest of capitalist Asia and began to feel that it was socialist central planning that had condemned it to backwardness and poverty.[44]

Trade Not Aid

Even in those parts of the developing world where foreign direct investment had been viewed with suspicion as exploitative and against a host nation's best interest,[45] there was a growing acknowledgment of the benefits that the free market and the opening up of the economy could bring. Frustrated with the poor yields of closed economic policy and import substitution, and having witnessed the success of the Asian "tiger" economies—Singapore, Hong Kong, Taiwan, and South Korea—many other developing countries increasingly declared a willingness to open up their markets and welcome the doctrines of free market capitalism. They had seen how these countries had entered into licensing agreements and joint ventures and utilized the capital or the technology of foreign corporations and investors with notable success. Furthermore, with the collapse of the Communist trading bloc and the implementation of international trade treaties such as the GATT Uruguay round, which promoted liberalization and deregulation of global markets, developing countries had few options. By the middle of the 1990s there was only one game in town. Governments that were once wary of foreign capital became caught up in the worldwide race for export-oriented growth.

In the meantime aid, the traditional tool for development, was being steadily withdrawn by first world nations. In 1992 foreign direct investment (FDI) overtook aid for the first time, and the gap has continued to widen. In 1997 FDI in the developing world exceeded $160 billion, while official development flows (that is, aid) in that year reached a mere $40 billion; by contrast, in 1990 aid totalled almost $60 billion and FDI just over $20 billion. Rich countries, now mindful of their public expenditure, were cutting back on expenditure outside their immediate sphere and understandably preferred to cut back on aid rather than on services at home, which were more likely to bring in votes. In any case, the political motive for aid no longer existed. During the cold war era many developing countries were of strategic importance and aid had become a currency used by many developed countries to buy allegiance and compliance. Once the Communist threat was gone aid to "friendly" countries fell sharply. Nowhere was this change in policy more marked than in the United States. While in the late 1940s, 15 percent of every U.S. tax dollar was sent overseas under the Marshall

Plan, in 2001 American foreign aid represented less than one tenth of one percent of the government's $1.9 trillion budget.[46] Handing over money to other states was hardly in keeping with the new probusiness approach of Western nations. The state was, after all, now seen as a conduit for private sector growth rather than an engine of growth itself. Better to put money into the private sector in these countries wherever possible. Aid projects often became tied to private sector development.

Ideological Consensus

By the early 1990s the laissez-faire neo-liberal capitalism of Reagan and Thatcher had unquestionably become the dominant world ideology. Even the traditional left now embraces many of its key tenets.

We see this most clearly in the United States and the U.K., where much of the legacy of the Reagan–Thatcher era is now ineradicable. In the United States, for example, the Democratic Leadership Council—an influential group of modernizers within the Democratic Party—pushed the party away from Michael Dukakis's leftish position toward the center, reinventing it as the "New Democrats." In place of the former concern for social justice, the New Democrats, exemplified by Governor Bill Clinton, emphasized business, investment, competitiveness, and free trade and targeted tax credits rather than public spending increases.

This ideological realignment was paralleled by developments on the other side of the Atlantic. In Britain, for example, in 1994 the Labour Party, recovering from its fourth successive election defeat, made a decisive break with the past, abandoning its traditional tax-and-spend policies (which were widely seen as the primary reason for its defeat) and embracing neo-liberal free market economics. The new leader of the Labour opposition, Tony Blair, actually endorsed the framework created by Conservative Chancellor Nigel Lawson in the 1980s, saying that a Labour government would balance the budget and have an explicit target for low and stable inflation. Employment would be managed by supply-side policy.[47] Clause Four of the Labour Party's constitution, with its commitment to public ownership of the means of production, was dropped.

In both New Zealand and Australia in the 1980s, it was "Labour" administrations, not Conservative ones, that presided over the dismantling of the social democratic order and were the first architects of neo-liberal restructuring.

In a world governed by free market capitalism—for no other system has proven as effective in generating wealth—the traditional political spectrum defined decades ago by pro- and anticapitalists is no longer fitting. The new parties of the center-left no longer place themselves anywhere along a left-right axis. Blair has frequently spoken of the need to "move the political debate beyond the old boundaries between left and right altogether."[48] Clinton in his 1992 manifesto denounced "the brain-dead old parties of left and right."[49] Today Republicans and Democrats alike are free traders who back the North American Free Trade Agreement (NAFTA) and the role of the World Trade Organization (WTO). Neither party would want to raise taxes or public expenditure, or change current monetary policies. Neither would be prepared to starve the burgeoning private sector.

In newly emerging markets such as Eastern Europe and South Africa, thanks to the Washington consensus and the influence of private corporations on policy creation, the Anglo-American system has also gained primacy. In Eastern Europe, where explicitly left-wing parties are tarnished by the legacy of communism, there is no space for a constructive left-of-center opposition to economic liberalization. Outside the extremist parties, politics in these countries is premised on the assumption that free market capitalism is essential to prosperity. In South Africa the Marxist rhetoric of the ANC in the 1990s, which espoused redistribution, social and public spending, and welfare, had by the end of the millennium adopted the now standard Anglo-American line of fiscal and monetary conservatism, trade liberalization, and privatization.

Only Asia, parts of Latin America, and continental Europe have seemed reluctant to fully embrace the new consensus. Asian governments continued to intervene in the economy throughout the nineties. To their cost—as the free marketeers later claimed, blaming the Asian financial crisis and subsequent downturn on excessive government intervention, crony capitalism, and market inefficiencies. Subsequently, aid has been provided only in exchange for market reforms of the American ilk, and little attention has been paid to the fact that these countries are not at all like America, with significantly different cultures, levels of development and institutions, and very different needs.[50] In Latin America a new wave of populist presidents has been elected—Luis Inacio Lula Da Silva in Brazil, Hugo Chavez in Venezuela, Nestor Kirchner in Argentina, and Lucio Gutierrez in Ecuador—all of whom promise their constituents a more regulated and more redistributive form of capitalism than Washington offers.

The ruthlessness of the Anglo-American model never sat well with most continental European politicians, who still value the underlying principles of the social model—solidarity, achieved through comprehensive welfare systems and economic cooperation, and a belief that the economy should be regulated for the sake of society—and instinctively feel laissez-faire capitalism, with its emphasis on deregulation and privatization, to be excessive. They see the U.K. as a Trojan horse, infiltrating Europe with American probusiness ideology. However, concerns about increasingly ageing populations, unemployment pressures, and moves toward monetary union (the euro) in Europe have meant that even traditional European socialists are having to accept the predominant zeitgeist, to a certain degree at least, and are adopting policies that not long ago would have been seen as frankly heretical.

The official age of retirement for public sector workers has been raised in Germany, Greece, Italy, and Finland, while the levels of pensions have been reduced.[51] In France, Lionel Jospin's government, facing unemployment levels of 10.6 percent, began to reevaluate the disincentives for employers imposed by a relatively high minimum wage and extensive social security charges.[52] For the first time, Jospin's advisers talked about the problem of "poverty traps," whereby generous welfare checks discourage the unemployed from looking for a job. His economic team worked on "tax and benefit changes, along [British Chancellor Gordon] Brown's lines, to steer the jobless into work."[53] Jospin also promised tax cuts for the middle class.[54] Many other countries have introduced unemployment reforms to force recipients to accept work at market-driven rates.[55]

Fiscal disciplines imposed on the dozen or so countries that strove to join the single currency by the Maastricht Treaty have made center-left governments as fiscally rigorous as right-wing ones.[56] And in order to meet the convergence criteria, European Union member states had to adopt conservative macroeconomic policies and ensure that fluctuating levels of national debt, government spending, or interest rates were not creating major fluctuations in the value of the currency in one state, which the other member states would then have to bail out.

Most of Europe now acknowledges that the social model must be reformed in the interests of economic competitiveness. Increased competition from other countries for inward investment has forced all social market economies to buy in to the free market doctrine to some degree and deregulate aspects of their labor and capital markers, lower taxes, and shrink their welfare states in order to remain a contender in the eyes of increasingly portable global corporations.[57]

Even traditional left-of-center parties in the 1990s began advocating slimmer government, lower taxes, and privatization, measures to which they were once bitterly opposed.[58] By the end of the millennium Greece's Socialists were slashing state spending to try to squeeze the drachma into the euro. And France's Socialists had privatized more companies than their immediate right-wing predecessors had.[59]

In Berlin, Social Democrat Chancellor Gerhard Schröder has seemed the most willing of the continental European left to emulate Clinton and Blair by moving toward the right. While still arguing for the importance of social justice, the Blair–Schröder joint manifesto on "the third way" enthused about deregulated markets, entrepreneurship, cuts in taxes and public spending, and a minimalist state.[60] Since being elected in 1998, Schröder has proposed cuts in taxes on both corporations and personal incomes.[61] He has also tried to shrink the national debt, reduce public spending, and lower state pensions and other welfare benefits.[62]

Yet despite the fact that the European social model is clearly being forced to adapt, talk of its *death* is perhaps premature. Jospin, believing that he was saying "yes to a market economy, no to a market society," continued to intervene in the employment market, and kept taxes at high levels; Schröder still talks of "common good" and has now eased back from the Blairist talk of a *Neue Mitte* ("new middle way") in favor of "our German model"; and even in the U.K., Gordon Brown, the chancellor of the exchequer, acknowledged in February 2003 that the overextension of the market in the public sector could have "long-term irreversible and catastrophic consequences."

However, all European states have adopted *some* of the liberalizing policies characteristic of the new Anglo-Saxon model, and across Europe there have been moves toward capital deregulation, welfare reform, and privatization. At the beginning of the millennium it seems that, while not renouncing consensus politics and welfare spending as thoroughly as in Britain, continental Europeans are accepting more readily than before a reduction of the role of the state in guiding national economies, and increasingly believe that corporations and entrepreneurs can generate wealth more effectively than governments.[63]

We are witnessing the emergence of a new consensus, different in content, but similar in tone to that which prevailed prior to the 1970s. In a speech in 1968 Margaret Thatcher said: "There are dangers in consensus: it could be an attempt to satisfy people holding no particular views about anything. No great party can survive except on the basis of firm beliefs about what it wants to do."[64] The irony is that the firm beliefs of Thatcher and her contemporaries across the world may have robbed their successors of plausible alternatives. In the triumph of free market capitalism we are left with a single world ideology.

One World

While ideas have been converging, the world has been shrinking. The state has been stepping back, and the market has been taking over. The liberalization of international finance started in 1960 with the development of offshore lending, continued with the collapse of the Bretton Woods agreement in 1971 and the floating of major currencies, and was completed with the deregulation of the financial sector in the 1970s and early 1980s and the subsequent invention of new financial products: derivatives and options with ever more enticing names—butterflies, straddles, and the like.

Access to these new financial products has resulted in an explosion in capital flows, greatly helped by the recent revolutions in communications.[65] The cost of a three-minute telephone call between New York and London has fallen from three hundred dollars (in 1996 dollars) in 1930 to less than forty-five cents today; the cost of computer processing power has been falling by an average of 30 percent a year in real terms over the past couple of decades.[66] Billions of dollars are being transferred all over the globe in real time every hour of the day by institutional investors and mutual and pension funds, more willing and able than ever before to diversify risk by putting their money abroad or moving it from one place to another.

Not since the end of the nineteenth century have we seen such an outflow of overseas investment[67] with governments once again increasingly powerless to control or contain these cross-border movements, as clearly shown by the Long Term Capital Management (LTCM) fiasco in which the American Federal Reserve ended up having to coordinate a bailout of the LTCM fund, only one among many severe financial crises in the late 1990s.[68]

But it is not just portfolio investment that has been on the rise. Since the early 1980s, corporations, in their quest to secure more cost-effective manufacturing bases and reach new markets in an ever more competitive business environment, have been not only exporting more and more—world exports more than doubled from 1980 to 1998—but have also been investing in overseas operations and setting up subsidiaries at an unprecedented rate, aided by the increased mobility of international capital, which allows them to raise money in offshore locations and move it across exchanges; by the communications revolution, which means that the head office can now easily communicate with its subsidiaries anywhere in the world; by the cost of transportation, which has plummeted especially relative to the value of traded goods; and by the increasing openness of markets that hitherto were closed.[69] Rich countries' industrial tariffs average now only about 4 percent.

Corporations think nothing nowadays of breaking up their chains of production and locating the links all over the world wherever it seems most advantageous. Designing their products in one place; entering into production alliances in another; outsourcing components and service activities somewhere else; sourcing their inputs, capital, raw materials, and even labor from wherever costs of production are lower, tax benefits more favorable, and access to raw materials or skills cheaper; and marketing in yet another place. When GE, for example, wanted extra cost savings on appliances, turbines, and jet engines, it told its American suppliers to move their operations to Mexico or elsewhere where labor was much cheaper, or it would find different suppliers.[70] Even firms previously comfortably situated within their home territories and relatively domestically oriented have recently disembedded their production and main operations from the parent state, seeking to lower production costs and expand in developing markets.

Multinational corporations, fed to bursting by global laissez-faire capitalism, are now as big as many nation states. Three hundred multinational corporations now account for 25 percent of the world's assets. The annual values of sales of each of the six largest transnational corporations, varying between $111 and $126 billion, are now exceeded by the GDPs of only twenty-one nation states.

Corporate sales account for two thirds of world trade and a third of world output (Coca-Cola, Toyota, and Ford derive nearly half of their revenues outside their base in the United States), while as much as 40 percent of world trade now occurs within multinational corporations. And these corporations are not only selling globally, they are investing all over the globe. Foreign direct investment has exploded, increasing from around $60 billion in 1980[71] to over $700 billion in 2001,[72] with its impact most noticeable in the developing world. While developing countries received an average of $2.35 billion in the early 1970s, this had increased to $80 billion in the period of 1991–1996.

Identical products are now manufactured for distribution all over the world. Brands have become universally recognizable, their attributes signaled in the international language of advertisements on far-flung satellite television screens. Coca-Cola has been imbued with so much value that it has become the traditional drink at Indian weddings; the blue, red, and white of a Pepsi can is now more identifiable than the Union Jack; the swoosh insignia on a pair of Nikes, as familiar in Milan and London as in Saigon, has spawned an entire new global industry of pirated swooshes.

I Want to Be a Millionaire

Globalization has not only created a plethora of choice but also a convergence of aspirations and values, which now center around people's desire to own, acquire and—as Adam Smith put it—"truck and barter." From New York to Moscow, from Bhutan to Borneo, we all increasingly covet the same products, the same brands. The success of the game show *Who Wants to Be a Millionaire?*, which is now shown in fifty-one countries, garnering regular audiences of up to 20 million, shows how much we all want to share the capitalist dream. No longer content with just watching the rich get richer—*Dallas* and *Dynasty* lost their appeal by the 1990s—today we like to believe that wealth is within reach of all of us.

And is it? What is the net result of global capitalism, of a world in which people's economic well-being and physical safety are determined primarily by the strategies and actions of international financial investors and multinational corporations? A world in which the primary service that national governments appear able to offer their citizens is to provide an attractive environment for corporations or international financial investors?

There has been an unprecedented rate of growth of material prosperity not only in industrially developed nations but in countries that, at the close of World War II, were part of the impoverished third world.[73]

In the United States, before the current recession took hold, the new century began during the longest period of economic growth in its entire history, with the lowest unemployment rates in thirty years, and the first back-to-back budget surpluses in forty-two years.[74] American corporations experienced remarkable growth rates in the nineties, and individual CEOs were amply rewarded for shepherding the boom. Sandy Weill, the head of Citicorp, received $200 million in compensation for 2001. Michael Eisner, head of Disney, earned $576 million, or roughly the GDP of the Seychelles.[75] CEO share option schemes rose over the decade from $60 billion to $600 billion.

In Britain the proportion of the population who owned their homes rose from just over a half in 1980 to two thirds by the end of the Thatcher period.[76] For many, standards of living increased and sales of TVs, compact discs, freezers, and cars all went up.[77] Four in five households in the U.K. now have a video recorder, and 34 percent of U.K. households have a home computer, this latter figure having almost doubled in the past two years. Sixty percent of the population now call themselves middle class, and that includes half of those in skilled manual occupations. Unemployment is at the lowest levels since 1980. Inflation persists but at a tiny fraction of the levels of the 1970s. More of us have more money to spend than ever before, and more places to spend it. Europe's largest multiplex cinema has just been built in Birmingham, England, and the British public now spends more on leisure activities than on food, housing, or clothing. Overseas holidays are commonplace: British residents took 56 million holidays of four nights or more in 1998, up a third since 1971.[78]

In New Zealand, growth since 1992 has averaged 4 percent a year, and unemployment has almost halved, to 6 percent.[79] Australia is enjoying one of the highest growth rates in the developed world. Chile experienced a decade of annual average growth at 7 percent from 1988 to 1998.[80]

Much of the third world has also experienced an investment boom. Net private capital flows to developing countries are six times greater than in 1990, and foreign direct investment has greatly raised the absolute economic welfare of most host countries through tax revenues to government. Royal Dutch Shell, for example, active in over seventy-five developing countries, in 1998 generated tax revenue globally of more than $46 billion.

Several countries that have opened their arms to free market principles have seen this strategy pay off. Singapore, albeit with an authoritarian regime, has very low unemployment, high single-digit growth rates in GDP and GNP, a highly skilled workforce, a 91 percent literacy rate, and a per capita income that is the second highest in the region after Japan. Thailand has tripled its GDP per capita since 1975 when only one in six people in rural areas had access to safe drinking water; today it is four out of five.[81] In India, which in recent years has relaxed its hostility to foreign investment and liberalization, the economy is booming; car sales in cities jumped by 57 percent during the first nine months of 2000, and Indian software developers are making a global impact, with software sales grossing approximately $4 billion in 2000. Along the northern border of Mexico, where the Maquilla export zone has been set up following the establishment of NAFTA in 1994, multinational production created over half a million new jobs where virtually none existed before, often providing better benefits and paying high wages than local companies.

Even in China, where free market reforms have taken place within a nominally Communist state, "liberalizing reforms led to a doubling of grain production in five years and provided a new demonstration of the power of market principles,"[82] and foreign investment has provided training opportunities, upgraded local facilities, and created countless new jobs wherever it has been applied. Companies do not tend to lower their standards in their foreign operations, at least not consistently as one might have thought. Often they improve local operating standards by exporting their own rather than conforming to local norms. When the Polaroid Corporation opened its plant in Shanghai in 1990, it brought engineers from its Scottish facilities to reconstruct the same basic plant design and working conditions as in Scotland. "The only notable differences were the installation of a high-end stereo system in the plant's main assembly room and an in-house clinic (customary in

China, but not in Scotland)."[83] Standardization is more important to many firms than scrambling to wring cost advantage from local loopholes or regulatory gaps.

Laissez-faire capitalism appears to have triumphed. Handing over the economy to the market seems to have been the right choice. It all looks great. At first glance, that is. But as they say, there is no such thing as a free lunch. So what is the price we will have to pay?

3

Let Them Eat Cake

The Prince and the Pauper

On December 4, 1998, Prince Charles paid a visit to the London offices of the *Big Issue*—a magazine sold by those who have fallen on hard times and who benefit financially from their sales—where he came face-to-face with homeless man Clive Harold, one of the magazine's street sellers. He astonished both the prince and the press by greeting Charles warmly with the words, "Remember me?"

Harold, dressed in a Father Christmas hat and burly, shabby overcoat, reminded the prince that both had been in Form Five of Hill House School in Chelsea, London, in 1957 when they were nine years old. Their paths had diverged dramatically since then.

The son of a wealthy London financier, Harold had gone from a childhood living in a five-story house in exclusive Launceston Place, Kensington, to sleeping outside in the Strand. Now he was living on welfare in a twelve-pound-a-night bed-and-breakfast hostel, trying to get his life back on track by selling the *Big Issue* outside Holborn Underground station.

He and Charles had parted ways at the end of the 1956–1957 academic year. Charles went to Cheam and then Gordonstoun school, Harold to Millfield, the most expensive private school in the U.K.

By the seventies Harold had become a successful journalist. He had been a reporter for the *Sun* and in 1980 published a book on UFOs, which rose to number eight on the bestseller list. He had been twice married, with three children, by his two wives and a live-in lover. A long-term heavy drinker, Harold's problems stemmed from 1987, when both his father and stepmother died in the same week. He descended into alcoholism, blew the £30,000 inheritance which he received in 1991, and lost his house. "One day I woke up in a doorway in the Strand," said Harold. "I thought to myself, What the hell have you done?" He had been sleeping in the Strand until three months before the meeting with Prince Charles—this after a successful, even glamorous life with frequent business trips to Los Angeles and New York, good hotels, all the trimmings.

But the late eighties and the nineties had not been kind to him. Frequent attempts to kick alcohol and revitalize his career had not been successful. After the meeting with the prince, he said: "Now I'm getting my confidence back. With the help of the *Big Issue,* I'm on benefit and living at a bed-and-breakfast place. I've joined a writing class and teach others to do what I should be doing."

The prince commented to the press: "Even with a supportive home background young people today can find it hard to maintain their self-confidence against the enormous pressures of modern life. My meeting with Clive Harold was a vivid reminder that homelessness can happen to almost anyone. We live in an increasingly materialist and secular world in which people's identity is determined so often only by the job they do and the money they earn, rather than by what they contribute to society as a whole."

John Bird, founder of the *Big Issue,* said: "We've had millionaires' sons, Old Etonians and army officers in here. Anyone can find themselves on the streets, no matter what start they had in life." At the time of their meeting, Harold was making £150 a week selling the *Big Issue.* Charles, by contrast, was valued at £100 million.

Harold sold his story to the *Sun* for an undisclosed sum. His former wife Eva and ten-year-old daughter then gave their side of the story to the tabloids, detailing Harold's neglect of his family and the two-week stretch he had done in Pentonville for failing to pay maintenance. For more than a week, he was headline news. A year later, he was still selling the *Big Issue.* The magazine's head office said that he was so traumatized by the press attention that he had asked the company to preserve his whereabouts and anonymity at all costs.

Things do go wrong. Not everyone benefits from the capitalist dream.

The twenty-year neo-liberal experiment that began in Westminster and Washington has not delivered for all of us. Traditional measures of economic growth, such as GDP per capita or GDP growth rate, obscure the truth. Not only has Clive Harold been excluded from the growth process, countless others are being excluded, too.

East-West, North-South

The global policy shift toward neo-liberalism that took place during the 1980s and 1990s was supposed, according to its proponents, to bring a convergence of living standards of richer and poorer nations. This never actually happened. For the majority of developing and transitional economies, the East-West and North-South income gaps are greater today than before.

The medicines doled out by the World Bank and IMF—"shock therapy," "stabilization," "structural adjustment," "trade and financial liberalization," "deregulation"—eroded labor institutions and diluted union bargaining power, led to rushed-through mass privatization programs that only benefited a minority, and prohibited countries from increasing public expenditure to meet their welfare needs. Not only was the pill bitter, it was often force-fed. The IMF and World Bank can dictate terms to the developing countries that depend upon loans from the international community, by making their loans conditional on the acceptance of their economic views by these nations. Through financial dependence or the threat of sanctions, these organizations coerce errant states into compliance.

Often the patient is made worse. Inequality declined *within* many countries between 1945 and the 1970s, but since the tenets of the Washington consensus became mainstream, there has been a reversal of this trend all over the world.[1] In the whole of the former Soviet bloc, most of Latin America, and much of South, Southeast, and East Asia, inequality has risen significantly over the past two decades. For example, 75 percent of the Mexican population live today in poverty, up from 49 percent in 1981.[2] And "with the notable exception of East Asia, the number of people living in extreme poverty—considered here as living on less than a dollar a day—has increased over this period in every developing country in the world."[3]

Even countries that assiduously followed Washington's dictates are not benefiting. Since 1994 South Africa has pursued a policy of seeking close integration into the global economy. Trade liberalization in some sectors has proceeded even faster than GATT/WTO requirements. Effective state support for industries such as clothing and textiles, which are major employers of labor, has been rendered impossible. Exchange controls have been steadily abolished, restrictions and regulations in respect of foreign direct investment have been removed. Yet the rewards have been limited, and economic performance poor and patchy. Postapartheid South Africa has suffered from slow growth, rising unemployment, and an alarmingly poor rate of delivery of social and physical infrastructure.[4] The top 10 percent of households account for almost 50 percent of total consumer spending while the bottom 10 percent account for just over 1 percent.[5] It seems likely that the nation's strictly market-driven industrial policy will undermine rather than promote social reconstruction.

This is a lesson China would do well to heed, now that it has decided to join the WTO. On the terms finally agreed with the United States and the EU, not only must China quickly liberalize its trade policy, it must also completely dismantle the apparatus of industrial policy that provided support to its state-owned enterprises. Given that most of these firms are not viable in an unprotected, unsubsidized environment, if China does honor the terms of the agreement the implications are likely to be great. The jobs of 90 million people are potentially at risk, and there is no social safety net to cope with such losses.[6]

Even in those third world countries that have been experiencing higher levels of aggregate growth through the embracing of neoliberal economics, for example, Chile, the money that is being made is not being distributed among the population. Only a minority benefit from the gains.[7]

Multinational companies that are now able to operate in developing countries as a result of liberalization policies are among the increasingly few that do benefit. Attempts by third world governments to attract foreign investment, direct or portfolio—ever more urgent because of the dramatic cutback in aid flows over the past few years—often precipitate what has been called a "race to the bottom": they limit or dismantle regulation, lower wages, slash welfare requirements, and tacitly allow corporations to create huge social upheavals. Pension contributions have been scrapped, and health care paid for by employers reduced. Potentially disruptive groups such as organized labor, which may jeopardize the quest to attract and use foreign investment and expertise, have been silenced[8]—in China hundreds of trade unionists are in prison or labor camps simply for having tried to form unions in special economic zones for foreigners. Short-termism is the norm: "pollution havens" created as environmentally unfriendly policies are allowed far below socially desirable levels,[9] human rights abused, a blind eye turned to illegal acts, all in an attempt to attract foreign investment and all in the name of free market capitalism.[10]

But it is not only multinational corporations that benefit from "open door" policies. Other beneficiaries of such policies are, typically, the host government, corrupt officials, and those fortunate enough to gain employment with foreign firms—they tend to pay better and their standards are often higher than those of local firms. In most of the third world, those outside the ruling elite, or outside the factory gates, are excluded from any gains.

China is the country that has benefited most from the greatest amount of FDI over the past few years, and has had astounding year-on-year economic growth for the past twenty-odd years, yet over a fifth of the population lives on less than one dollar a day. India is the third world's software success story of the decade, yet around half of the population lives on the per-day equivalent of a dollar and a half. The divide between rural and urban China and rural and urban India is now so great that it doesn't even make sense to think of these two contrasting worlds as the same country.[11] And while 80 percent of revenues generated by oil companies operating in Nigeria remained in the country, they benefited a tiny ruling class.

Rather than the rising tide of the market lifting all boats, structural adjustment and liberalization policies with no concomitant obligations on redistribution appear to have sunk some social groups, especially the vulnerable and the poor.

A senior oil executive operating in Colombia said of his company's various activities there: "These projects are good for the government and good for us as a company . . . it's good for whoever can exploit the situation . . . but the majority don't benefit . . . it's only good for the few who can impose themselves."

In some places those who oppose the inflow of investment, fearing that too high a price will have to be paid, fall victim to state officials' greed. There are several reported incidents of Nigerian security forces beating, detaining, or killing people who protested against oil company activities. In November 1995 Ken Saro-Wiwa, the Nigerian political activist, poet, and playwright who had led a five-year battle for the secession of Ogoniland and for compensation from Royal Dutch Shell for the environmental damage it had wrought there, was one of nine Ogoni activists executed after a trial that clearly violated international fair trial standards.

In India in May 1997 police beat 180 protesters demonstrating peacefully outside the gate of the now discredited Enron Power Corporation against an electricity-generating project which locals feared would divert scarce water and kill fish.[12] Enron may not have controlled the chain of command, but the police clearly knew whose interests they were expected to protect. In the pursuit of foreign capital—often used to line their own pockets—some third world governments have shown themselves willing to literally sacrifice their citizens.

The injustice still seen today in the third world remains shocking, for all its familiarity. Conventional wisdom explains it as a product of the combination of economic underdevelopment and weak or absent democratic institutions. The conventional analysis is that in the long run, opening up a country's economy and encouraging inward investment will help to improve the people's lot, because it fosters education and training and so produces a better informed workforce and a reflective middle class—both of which, history suggests, are forces that work toward improved democratic participation and concern for rights and equality.

What the conventional analysis ignores, however, is that both the pursuit and the subsequent results of these free market policies can polarize the population to an increasingly unacceptable extent. This divisive trend can offset the more comfortable conventional prediction of what trade and investment will bring. The riots in Argentina in December 2001, for example, provide a stark testimony to the breakdown in social capital that arises from a situation of growing inequality: Two thousand additional people were falling below the poverty line in Argentina each day at this time.

Outside Your Front Door

This growing polarization is not just a third world phenomenon; we do not see it only in underdeveloped countries with weak democracies. It is also happening here in the West. Clive Harold is just one of many who have gained nothing from the recent boom years even from an initial position of considerable advantage. And the winner of the first million on the U.K.'s version of *Who Wants to Be a Millionaire?* was—you guessed it—already a millionaire: Judith Keppel, a distant cousin of Camilla Parker Bowles, Prince Charles's girlfriend.

For in the West, too, the gap between rich and poor is widening.[13] In America the spoils of a long period of prolonged economic expansion and low unemployment have not been widely distributed: 97 percent of the increase in income has gone to the top 20 percent of families over the past twenty years. While the rich earn more—average earnings of the top fifth of male earners rose by 4 percent between 1979 and 1996—the bottom fifth saw a 44 percent drop in earnings. And although the wages of blue-collar workers began to rise in real terms after the mid-nineties for the first time since the late seventies, middle American incomes continued to be held back by waves of corporate downsizing—creating a widening gulf between the middle and the top.

America is today the most unequal society in the industrialized West, with incomes now less equal than at any time since the Great Depression. The gap between the top and bottom 10 percent is so large that those at the bottom are considerably poorer than the bottom 10 percent in most other industrialized countries—the United States ranking nineteenth—even while the United States has the highest per capita incomes.[14] Some 36.5 million Americans (13.7 percent of the population) now live in poverty, while 40 percent of the country's wealth is owned by the top 1 percent, compared with 13 percent less than twenty-five years ago, meaning that the 13,000 richest households in America have almost as much income as the 20 million poorest households. While the national unemployment rate in the United States is 5.4 percent, on many of its Native American reservations the rate is as high as 70 percent. In isolated rural areas in America, the unemployment rate is often two, sometimes four, times as high as the national average.[15] And social security for the unemployed has become much more conditional: Only 39 percent of unemployed Americans have access to unemployment benefits today, compared to 70 percent in 1986.[16]

Under Bush the gap between rich and poor will only increase. His plans for the eventual elimination of both capital gains and inheritance taxes will create an aristocracy of the wealthy and condemn the rest to underdog status. As the legendary investor Warren Buffet says, just as it would be absurd to select the U.S. Olympics team for 2020 from the children of the winners of the Olympics in 2000, so it is wrong to construct a society whose likely leaders tomorrow—given the advantages that wealth confers—will be the children of today's wealthy. This offends not merely the values of democracy and equality of opportunity on which the United States is constructed, but will also be economically disastrous.

In the U.K., after eighteen years of Conservative rule, the situation New Labour inherited was similar if less marked. The distribution of incomes in the U.K. is more unequal now than at any time since World War II. When Margaret Thatcher came into power in 1979, the richest fifth of the population accounted for 43 percent of all earned income, and the poorest fifth 2.4 percent. In 1996, the last year of the Conservative government, the figures were 50 percent and 2.6 percent, respectively. Since in the same period Britain's GDP rose substantially, the poor are getting a smaller slice of a much bigger pie.[17] The number of families below the poverty line rose by 60 percent in the 1980s, and by 1996 the U.K. had the highest proportion in Europe of children living in poverty, with 300,000 British children worse off in absolute terms in 1995–1996 than in 1979.

The number of people living below the poverty line has continued to rise under New Labour, despite efforts to reverse the trend. The number of households existing on less than half the average weekly income of £278 ($417) after housing costs rose by 1.3 million to 14.25 million between 1994 and 2000, more than double the number in the early 1980s.[18] Some 500,000 of that increase occurred after Labour took power in 1997. Poverty is heavily concentrated among single-parent families (at least half the single-parent households with children have incomes below the poverty line) and in homes where nobody has a job. The proportion of pensioners living on less than 40 percent of average income rose from 20 to 23 percent between 1998 and 1999.

In this new competitive world of free market capitalism, it is the unskilled who fare worst. They have become the Epsilons of our new Brave New World—effectively commodities, easily replaceable by an ever-growing overseas supply; and, in our postmanufacturing era, in ever less demand.

Those who were already earning lower wages are earning even less: They have less and less political and economic clout, thanks to their diminishing attractiveness as a group now that technological advances have lowered the demand for the unskilled, the decline of union power, and the increased competition from lower-cost manufacturing outposts or migrant labor willing to work for significantly less. Jobs and incomes in rich and poor countries have become more precarious as the pressures of global competition have led countries and employers to adopt more flexible labor policies, and work arrangements that absolve employers from long-term commitment to employees. More than three fifths of American employers offer no form of employment contract; while over half of those that do include wording specifying that the employment can be terminated summarily for any reason.[19] The Netherlands, Spain, and the U.K. have all decentralized wage bargaining.[20] And even France and Germany have weakened their worker dismissal laws.

Those at the top of the food chain are being courted ever more assiduously— wages for top managers have been rising and perks for the "right" hires have been becoming even more extravagant,[21] with some of the biggest names in the recent corporate scandals being the most amply rewarded. John Rigas, CEO of Adelphia Communications, was provided with a $3.1 billion loan from his company. Dennis Kozlowski, CEO of Tyco International, made nearly $467 million in salary, bonuses, and stock during his four-year tenure at Tyco, a tenure which ended with the destruction of around $92 billion in Tyco's market share. And Kenneth Lay, CEO of Enron, was given $100 million in cash payments in a reign which ended with the forced redundancies of five thousand employees, the loss of around $1 billion from Enron workers' pension plans, and the wiping out of $67 billion worth of shareholder funds. This at a time when, in the United States, there has been a 28 percent decline since 1973 in entry-level wages in real dollars for male high school graduates and a decline in wages and benefits for all unskilled labor.[22] A fifth of American employees work at rates below the official poverty level, making a mockery of the low official unemployment rates, and even workers carrying out hazardous jobs are being paid less. In the late 1980s the union wage for removing asbestos insulation from old buildings was thirty-one dollars an hour, but by the 1990s the rate had collapsed, thanks to the rise in nonunion removal companies and an influx of immigrants eager for work. "Contractors had no trouble getting workers for twelve to fifteen dollars an hour—and workers willing to do the job without respirators," said Pawel Kedizor, the business manager for Local 78 of the Asbestos, Lead and Hazardous Waste Laborers Union.

Now increasingly able to operate globally, corporations "bottom fish," moving to countries with low labor costs to produce their goods. Production continues to shift to lower-cost options: from U.S. factories to the maquiladora plants on the Mexican border where nearly a million people are employed at wages of under five dollars a day; from Israel to nearby Jordan, putting scores of Israeli Arabs and Druzes out of employment as a consequence; from Silicon Valley to India and the former Soviet Union, where software is developed for a fraction of the cost it would entail domestically; from unionized workplaces to regions or countries where unions are less militant or there are nonunionized laborers, just glad to get a job.

And these jobs, even at reduced rates and with their spartan packages, are not even secure. For many, job security has become increasingly rare. Jobs are increasingly part-time, casual, contractual for those who still are in employment. In Latin America, for example, by 1996 the proportion of workers without contracts increased to 30 percent in Chile, 36 percent in Argentina, 39 percent in Colombia, and 41 percent in Peru.[23]

Technological advances have allowed machines to replace people. The "knowledge economy" requires less manpower. While the world's five hundred largest multinational corporations have grown sevenfold in sales, the worldwide employment of these global firms has remained virtually flat since the early 1970s, hovering at around 26 million people.[24]

Increased competition due to liberalization of trade policies has meant that inefficient industries have had to downsize or streamline (euphemisms, of course, for firing staff) or be forced out of business altogether. Even before the current recession, companies that were performing at levels that in the past would have been considered acceptable were firing staff, not because they were struggling but because the pressure on companies to make high returns was unprecedented (in many industries, returns of between 20 and 35 percent are now expected, with institutional investors today on average turning over 40 percent of their portfolio in a year, looking for higher returns), competition for investment flows was ever greater, and corporations felt more vulnerable than ever to threats of takeover or acquisition. Some 39 million Americans between 1980 and 1995 were caught up in a corporate downsizing program.[25] IBM fired 122,000 people between 1991 and 1995 and reduced total wages by a third in a bid to push up their dividends and share price. The return for such "prudence"? In 1995 the company's share price and dividend beat all previous records. The announcement by the American food company ConAgra that it would lay off 6,500 employees and close down twenty-nine of its plants pushed the price of its shares up so steeply that the company's market capitalization increased by $500 million in twenty-four hours. "For shareholders and managers, downsizing does pay off. Wall Street now simply prefers a dollar saved in costs to an extra dollar earned."[26]

Even those who have jobs are losing benefits. In the United States, where people are mostly dependent on corporations for health benefits and pensions, the consequences are particularly worrying. The 20 percent of Americans now working on temporary contracts or part-time receive no benefits at all, or insignificant ones. And while 70 percent of American workers have pension plans, less than 10 percent of those in the bottom tenth can rely upon any employer-financed retirement benefits. Compare this to the retirement package of Jack Welch, former CEO of General Electric, which, when he gave it up, included the lifetime use of a company jet, an $80,000-a-month Central Park apartment, membership at an array of country clubs, maid service at his multiple homes, flowers, phones, computers, furniture, limos, and prime tickets to the U.S. Open, Wimbledon, the opera, and every New York Knicks home game!

Add to this the problems engendered by the privatization of fundamental public goods and the situation is even bleaker. In the United States there have been numerous cases of HMOs having been exposed over the past decade as "cherry picking," that is, making sure that they attract healthy people and avoiding those who will be heavy utilizers of services. And most HMOs that were set up to take care of Medicare patients are now going bankrupt and getting out of the business since they cannot keep up with the escalating costs, especially prescription costs, of elderly patients.

But it is not just HMOs that are the problem. The ability to secure health insurance at all is proving elusive for significant numbers of Americans. Forty-five million Americans currently do not have health insurance, 25 percent of the chronically ill do not have adequate coverage. Disqualified from some insurance plans because of preexisting health conditions, viewed as high risk by others, and facing premiums they cannot afford, millions of Americans are facing potential crises. James Huth, a fifty-five-year-old with a heart condition and diabetes, on a monthly pension of $1,045, clearly cannot afford the coverage of $1,200 a month he has been offered. "I have a choice," he said. "Do I want to eat and have a place to sleep, or pay for health insurance and sleep in the street?" With Britain looking to America for answers to its health care problems and private health insurance being proposed as a solution, the American experience needs to be taken heed of. Given the track record of one of Britain's largest health insurers, PPP, which has refused to provide long-term treatment to a number of patients with chronic illnesses such as hepatitis C,[27] are society's needs and the individual's right to health care to be determined by some kind of actuarial process? Will people get cast on a scrap heap if companies deem their custom no longer worth it?

Even those succeeding well in this new world—those with good careers and prospects, health benefits and private insurance, those who fall into the top percentiles—are suffering. Stress-related illnesses, obesity, and diabetes are on the rise. The antidepressant market grew 16 percent per year in G7 countries between 1989 and 1999.[28] Sales of Prozac have eclipsed the GDP of small nations. Illness is now costing U.K. employers $17 billion a year. The seemingly unstoppable craving for wealth, for the trophies of capitalism displayed on every billboard but ever harder to bag, the perceived need for even bigger houses or cars—the average new American house, for example is now 2,200 square feet compared to 1,500 square feet in 1970[29]—are destroying the very fabric of people's lives. By mid-2001 the total stock of personal debt had climbed in the United States to a record 120 percent of personal income. And 1.5 million Americans filed for personal bankruptcy in 2002; this is 1 out of every 69 U.S. households.

Those with work are working longer and longer hours, with Americans putting in the longest working hours (around fifty hours a week)[30] among industrialized nations,[31] this despite the fact that they have less parental leave, less affordable day care, and the least number of paid holidays and vacations of all industrialized nations. British men emulating their American peers now work the longest hours in Europe: half of the U.K.'s fathers spend less than five minutes a day in direct contact with their children.

Having a life now is being traded off for the prospect of getting a significantly better life in the future. The Internet mogul wannabes who networked until the bubble burst in London at Home House, in Tel Aviv at Espresso Bar, and in San Francisco at the Thirsty Bear, ceded their relationships, their passions, and any semblance of a normal existence in the hope of becoming the next Larry Page or Jeff Bezos.[32]

In 1991 Ichiro Oshima, a twenty-four-year-old employee of Japan's Dentsu Inc., the world's biggest advertising agency, hanged himself after an eighteen-month career of eighty-hour weeks, days when he did not leave the office until 6:00 A.M. and then only to get changed before returning. Rare were the times that he managed to catch more than two hours' sleep. "I cannot function anymore as a human being," he told his superior about a month before his death. Japan's Supreme Court ruled that the company was fully liable for working the young man so hard he killed himself and for failing to prevent his death. And Ichiro's story was not an isolated incident: Suicides from overwork are becoming more frequent in a country which now even has a word for the phenomenon of people who work until they drop—*karoshi.* It is thought that there are ten thousand such deaths a year. Some fifty similar lawsuits are now before the courts.[33]

The West's booming economy in recent years has not ended homelessness, poverty, or inequity. Unchecked, the situation is only likely to get worse. Research has repeatedly shown that it is not those who live in the richest societies, but those who live in societies with the most egalitarian wealth distribution, that have the best health. It is relative income levels that matter, not as one might have thought, absolute ones. Death rates from some of the most significant diseases[34] are reduced when income differentials are lowered.

Similarly, higher crime rates, including murders and violent crime, are correlated with wider income differences,[35] both in the West and in developing nations, which make the statistics that 3 percent of the American male workforce is in prison[36] and that America's jail population has increased 800 percent over the past thirty years[37] at least comprehensible, if no less shocking.[38]

The effects of a society that includes increasingly marginalized groups cannot be escaped by moving into a gated community (although 3 million Americans have tried to do so),[39] or merely stepping over the homeless person who begs in front of your local store. The costs of this breakdown in solidarity—of ignoring the predicament of others, of perpetuating the selfishness born of a feeling of abandonment by the state—is likely to be carried by all of us.

By making economic success an end rather than a means to other ends, governments and people have lost sight of the fact that economic growth was supposed to have a higher purpose—stability, increased standards of living, increased social cohesion for *all,* without exclusion.

Things Are Only Getting Harder . . .

The advocates of global capitalism would argue that this is a temporary situation, a consequence not of flaws in the system, but of the fact that this experiment has not been running for long enough, that wealth will ultimately trickle down to all, "that by providing incentives to the rich, for example, through lower taxation, the rich will be spurred into entrepreneurial activity, which will in turn create jobs and boost growth," that free enterprise and free and open markets will ultimately deliver.

They might admit that the system is inherently ruthless, but would conclude that for societies as a whole, the costs of neoliberalism are worth paying. Thomas Mayer, chief economist in the Frankfurt offices of investment bank Goldman Sachs, has said: "When I look out on to the trading floor, I see no need to get the wages of the traders down. That's besides the point. But the need is to get the cleaning workers' wages down and to widen the spread between them. If we do that, we get more employment, less tax burden on those who are financing the unemployment, and therefore, greater growth."[40]

But who will look after capitalism's initial losers while they are waiting for the benefits to trickle down?

Not, it appears, most governments. The Reagan-Thatcher ethos is widespread, regardless of party persuasion. New Labour's Stephen Byers, in his first speech as trade and industry secretary in 1999 declared, "Wealth creation is now more important than wealth distribution."[41] And George W. Bush has announced a range of policies intended to take the enforcement of social justice out of the government's hands and put it into those of the community. Under Bush's Social Security plans, for example, every individual will provide for his or her own retirement with monies invested in stocks, thus breaking the relationship between citizen and government that had endured since the New Deal of the thirties. Under his "faith-based initiative," the provision of social needs will increasingly be put into the hands of religious groups and charities.

But even if governments want to address issues of social justice and inequity, can they really do anything significant? Seemingly little. As the former president of the German Bundesbank, Hans Tietmeyer, has said, "Politicians have to understand that they are now under the control of the financial markets and not, any longer, of national debates."[42] And if the financial markets deem that a new national health care scheme or a massive education reform will prove too costly, they will respond with higher interest rates or a collapsing currency. In this way global market forces not only rule out the kind of compensation to losers that would reduce the social costs of globalization, they also seem to challenge state sovereignty itself. The footloose, mobile nature of global capital increasingly dictates what governments can and cannot, singly at least, do.

Beggar Thy Neighbour

Pressure on governments comes not only from capital markets but also from corporations. The world of the twenty-first century is a seller's market for business. Advances in communications and technology and the deregulation of capital markets have meant that corporations are now increasingly portable, able to decamp and set up elsewhere with relative ease.

Recognizing the power that they now wield, multinational corporations play countries and politicians off against each other, exacting for themselves ever better and more lenient terms. Corporations effectively auction off promises of new jobs, infrastructure investment, and economic growth to the highest international bidder, declining to move to or threatening to pull out of countries whose employment costs and taxes are too high, or where standards are too stringent or subsidies and loans not forthcoming. Globally, dominant companies increasingly call the shots, able to move money freely, deciding for themselves where to invest and produce, where to pay taxes, and playing these potential sites off against one another.[43] Politicians are left trying to stem the flow, offering sweeteners to corporations to maintain factories so as to minimize the political and social costs of closure, but without any long-term guarantees that the firms will not eventually relocate. National governments appear increasingly impotent in the face of the giant corporations, who transcended national borders many years ago.[44]

The levying of taxes, arguably the most fundamental right of the nation state and a potential means of redressing social and economic inequality, is being squeezed by corporate pressure.[45] As capital and highly paid labor are now able to move more freely from high-tax countries to low-tax ones, as the world becomes more integrated in the wake of globalization and developments in communications, a nation's ability to set tax rates higher than other nations is being put in question. The resultant mind-set is one of "beggar thy neighbor." Ireland opposes harmonizing corporate tax rates across the EU because its low rates give it an advantage over other member states in attracting multinational firms. Britain blocks an EU savings tax directive because it might hurt the City of London. And corporate tax rates are pushed down the world over: The rates of U.S. affiliates operating in developing countries, for example, dropped from 54 percent to 28 percent between 1983 and 1996,[46] while in developed countries corporate tax rates fell from an average in the early 80s of 50 precent to under 35 percent in 2001.

In Germany, where revenue from corporate taxes has fallen by 50 percent over the past twenty years, despite a rise in corporate profits of 90 percent,[47] Finance Minister Oskar Lafontaine's attempt to raise the tax burden on German firms in 1999 was thwarted by a group of companies, including Deutsche Bank (assets over $400 billion), Dresdner Bank, the insurance conglomerate Allianz, BMW, Daimler-Benz, and RWE (the German energy and industrial group), all of which threatened to move investment or factories to other countries if government policy did not suit them.

"It's a question of at least fourteen thousand jobs," threatened Dieter Schweer, a RWE spokesman. "If the investment position is no longer attractive, we will examine every possibility of switching our investments abroad."[48] Daimler-Benz proposed relocating to the USA, other companies threatened to stop buying government bonds and investing in the German economy.

In view of the power these corporations wield, their threats were taken seriously. So seriously that they were undoubtedly a major cause of Lafontaine's resignation. He remained defiant to the end: "The heart isn't traded on the stock market yet," he said as he left.[49] "Things will be different now," Bobo Hombach, Chancellor Schröder's closest aide, commented in response. "We have to move in a different direction. Gerhard Schröder will have different priorities, that's obvious."[50] If Oskar Lafontaine's resignation proved anything, it was that Schröder was willing to take the pressure from business very seriously. Within a few months Germany was planning corporate tax cuts.[51] But by March 2003 any moves Schröder had made on that front were not enough. Employment laws were still viewed as tough, and taxes were still seen as a deterrent to investment. Deutschebank, the pillar of the German business establishment, was, it was revealed, now secretly considering moving out of Germany. And Infinion, the leading German microchip maker, announced that it would move the company headquarters and administration outside of the country.

It is not just corporate tax that concerns corporations. Countries with high rates of personal taxation are coming under pressure from the international business community, too. Several large Swedish companies, including Ericsson, have threatened to leave their home country because of high income tax which, they claimed, made it hard to recruit highly skilled employees.[52] (Ericsson actually did fulfill its threat, moving several corporate and production functions abroad and opening a big headquarters in London in 1999.)

In the twenty-first century corporations are increasingly deciding how much tax to pay and where to pay it. The Internet is only likely to make it even harder for governments to collect taxes. A company can now locate in a low-tax haven, base its physical production facilities (where it may angle for subsidies) elsewhere, and sell to its customers from a virtual location—outside of the reach of governments.[53] And the greater the advances in communications, the more cases we are likely to see of corporations locating in one place and paying tax in another or even nowhere at all. Companies like the banking firm BCCI, which through a complex web of aliases managed to be registered nowhere for tax purposes, and Rupert Murdoch's News Corporation, which has earned profits of over $2.3 billion in Britain since 1987 but, as of 1999, paid no corporation tax there at all and no more than 6 percent tax worldwide,[54] may well become the norm rather than the exception.

What is the impact of this noncollection of taxes from corporations? At worst, such damage to tax systems could occur that governments became unable to meet the legitimate demands of their citizens for public services—in the United States, for example, federal expenditure on roads, schools, and universities fell as a proportion of GDP throughout the 1990s even under a Democratic president. At best, the government's ability to spend is severely restricted, and an ever higher burden is put on individual taxpayers' shoulders. If governments were willing to implement such a redistributive strategy, that is. Usually, however, it appears they are not. George W's tax cut proposal was the centerpiece of his election campaign and remains one of his key tenets despite the rising U.S. budget deficit, the depletion of state revenues, the stalled economy, the cost of war, and the cost of the reconstruction of Iraq. And neither the Conservatives nor Labour in the 2001 British general election campaign risked campaigning on a tax increase platform.

Not only are tax-raising policies *perceived* as vote losers, the fear is that the rich or skilled would simply leave if they perceived the tax burden as too high. As able people of all kinds become increasingly migratory, it will be harder still to sustain heavy demands on them to bear the cost of social spending. Those with large incomes and capital resources are likely to follow the example of corporations, and relocate themselves with low taxes in mind.

As has Laetitia Casta, the twenty-one-year-old French actress, sometime Victoria's Secret lingerie model, and latest incarnation of Marianne, the personification of France, whose face will adorn French coins and stamps for the next decade. In April 2000 she moved to London, a relocation that made the front pages of newspapers on both sides of the Channel. Although Mlle. Casta has denied that taxation is her reason for moving, the predominant feeling in France is, as one French *député* has put it, that Marianne has moved to *perfide Albion* to escape the punitive wealth taxes of *la belle France.* With upper-level income tax rates of 47 percent in France, compared to 35 percent in the U.K., it seems that she had a very good reason for moving. More of her well-heeled compatriots are reportedly crossing the Channel to establish fiscal residence in one of the EU's most lenient tax regimes.

Despite the substantial differences in taxation between the U.K. and, say, Germany and France, and despite the relocation of Laetitia Casta and others, we haven't as yet actually seen significant levels of wealthy economic migrants to the U.K. Which suggests that although the raising of taxes is perceived by most politicians as far too risky a strategy to openly endorse or execute, it may be that in practice, doing so would not lose votes—for example, polls in the United States suggest that tax cuts are not of that much interest to most of the electorate.[55] And in the U.K. a large majority of the British public say that they are willing to pay higher levels of tax to improve public services. While the perception remains, however, a growing gap is threatened between the demand for public services and the taxation that is available to pay for them. It will be impossible to extend Medicare to include prescription drugs, for example, without cutting into the surplus the Bush administration claims is available for tax cuts.

Caring for the Corporation

Not only are governments finding it harder to raise taxes, they are also finding themselves having to provide "welfare" to a not terribly needy client—the private sector. In America direct subsidies to businesses total over $75 billion annually,[56] with the poorest states—those where the difference in income between rich and poor is greatest—offering the largest amounts.[57] In Ohio $2.1 billion in business property went untaxed in 1996 thanks to business assistance programs. Beneficiaries of Ohio's "corporate welfare" included Spiegel, Wal-Mart, and Consolidated Stores Corporation, all of which were absolved from property taxes. In the United States it is property taxes that fund public schools. As one school treasurer put it, "Kids get hurt and stockholders get rich."

Borden Chemicals has been excused approximately $15 million in tax over the past decade as part of Louisiana's corporate welfare program. The same Borden Chemicals that in August 1996 released a dense gray cloud of toxic ethylene dichloride, vinyl chloride monomer, and hydrogen chloride into the atmosphere; that in 1997 was responsible for the release of vinyl chloride monomer and ammonia into the atmosphere, which forced the closing of Route 73; and that in July 1998 released a cloud of hydrochloric acid fumes, shutting down roads in the area.

In Arkansas the state government spent over $10 million building new infrastructure to lure Frito-Lay to Jonesboro, Arkansas, while a neighboring town, plagued by dirty water supplies and in need of a new water infrastructure, waited for ten years for the $750,000 it needed to remedy the situation.[58]

And New York, Connecticut, and New Jersey spend more than $2.5 billion each year bidding against one another for what are essentially the same jobs. A record $720 million package of tax breaks and subsidies was offered by New York's governor and mayor to try to lure the building of a six-story office tower from New Jersey to New York. New Jersey had already tried to lure the Big Board with a substantial package of its own.[59]

This touting for business is by no means unique to the United States. The Taiwanese TV tube maker Chungwha generated a bidding war between Scotland and Wales at the end of 1995. General Motors touted the positioning of a $750 million car plant all across Southeast Asia.[60] The pharmaceutical company Glaxo allegedly threatened to relocate its British research and production facilities if its new anti-influenza drug, Relenza, was not approved for sale in the U.K.[61] And Tony Blair supported efforts to persuade BMW not to shut Rover's Longbridge plant, granting a £150 million subsidy in order to save ten thousand jobs at Rover. Within a year, however, BMW had pulled out of the U.K., giving no notice of its plans. Although the granting of subsidies has become a vote-winning strategy, it is not always enough.

But when governments provide welfare handouts to corporations (sometimes even in defiance of their own regulations) and respond to the threats made by corporations to pull out and relocate, the rules of the game subtly change. It comes to make good business sense for corporations to turn their attention toward the political arena. Competitive advantage may be gained not only from lower costs, better service, or differentiated products, but also by a company's political strategies and effective lobbying.

Ordinary citizens once again lose. In 1999 state and local governments in the United States gave businesses over $1.7 billion in tax rebates and subsidies. If this money had been spent instead on schools it would have been enough to educate one and a half million elementary school students at double 1999's average rate per pupil.[62] Not to be sniffed at, given that U.S. public education is in such disrepair that students achieve among the lowest scores in international rankings for performance in math. Money spent on subsidizing corporations is money that becomes unavailable for public services.

Furthermore, the promised jobs and investment flows often do not materialize, or if they do can swiftly vanish (as at Rover), or even be withdrawn. AT&T, Bechtel, Boeing, General Electric, and McDonnell Douglas (now a part of Boeing) received 40 percent of all loans, grants, and long-term guarantees given out in the 1990s by the Export-Import Bank of the United States, which subsidizes companies that sell goods abroad, yet overall during this period employment fell by 38 percent and more than 300,000 jobs disappeared. Chase Manhattan, which received $234 million of incentives in 1989, has since cut thousands of jobs. In New York the state comptroller found that about half a sample of loan recipients from the IDA, a body that issues tax-free bonds to provide low-cost financing to firms looking to move to New York, had miserably failed to meet their job targets. Study after study reveals no statistical evidence that business incentives actually create jobs.

And companies often take the money and then relocate with relative ease to places where the terms are even sweeter, despite the damage they can wreak en route. In Germany workers would at least have to be consulted in such cases, but in the United States and the U.K. they typically don't.

As Ohio Senator Charles Horn has said, "We know companies are manipulative, but it's the nature of business to go after every dollar that's legally available. Don't place the blame on the company; place the blame on government. This is government's folly."

Quite. But unfortunately a folly that persists. In the aftermath of September 11 the very same multinationals that over the preceding decade had been more than happy to recast themselves as global and had mocked the values of public purpose—IBM, AIG, Boeing, General Motors, and International Paper—were to be found lining up in the corridors of the Capitol and the White House lobbying for tax cuts and the $100 billion economic stimulus bill, which included a repeal of the corporate alternative minimum tax. If repealed, it would mean in effect that some large companies would pay zero tax in perpetuity. "It's a sort of free-for-all," said James Albertine, president of the American League of Lobbyists. "It's like squirrels running around finding acorns and putting them in the ground for winter."[63]

Passing the Buck

So governments are caught between a rock and a hard place; unwilling to risk losing the electorate's or corporate sector's support by raising taxes, and unable to increase spending for fear of market censure. But what will be the impact of the enforced capping of government's social spending? More inequity? A world in which the poor become ever more marginalized, and the rich even richer? A world in which the principal division is between those who are inside and those who are outside the global corporation? A world in which the consumer has some power, but those who cannot afford to be consumers have none?

Politicians from all mainstream parties have over the past few years espoused noble aims of reconciling capitalism with humanity, social justice with economic success. In the United States and the U.K., Clinton and Blair, for example, talked of a "third way" while Bush spoke of "compassionate conservatism."

In the United States, at least, such aims sound increasingly hollow. The country regularly stands last among developed nations in the proportion of its GNP devoted to social programs or to redistribution. Bush's phasing out of the "death tax," which affects only the wealthiest 2 percent of the population with more than $1.35 million to leave to their children, and his $350 billion in proposed tax cuts over the next ten years in ways that will substantially benefit the better off in society, show very clearly the extent to which his administration is intent on favoring the rich. When asked to justify a policy that under Bush's initial version would have provided the top 1 percent with a 43 percent share of the cut, Bush's answer was always the same: the 20 percent who will receive the biggest tax cuts are those who most deserve it, because they pay 80 percent of the nation's taxes. This at a time when the U.S. Census Bureau estimates that 3.7 million American households suffer from hunger as a result of being unable to afford to buy basic food items; and many more, about 9 million households, have "uncertain access to food."

In the U.K., New Labour has made some inroads into inequity while operating within the confines of the global capitalist system. Despite the discouraging results so far in tackling poverty, the poorest households have on average gained from recent budgets with the bottom tenth 9 percent better off, against virtually no gain for the top tenth. More than £2.5 billion in spending has been shifted away from the top half of the income distribution to the bottom, and various national measures have been introduced to reduce inequality of opportunity. And in July 2000 Chancellor Gordon Brown indicated his commitment to break what had almost become a taboo, increasing public spending with a £43 billion package which would put more money into education and employment, child care, transport, poorer regions, and inner cities over the next few years. New Labour has not been honest, however, about the trade-offs that will undoubtedly have to be made if this goal is to be pursued in the long term. It is only thanks to the strong economy, which resulted in falling payments to the unemployed and booming tax receipts; to one-off moneyspinners, such as the auction of mobile-phone wavebands; and to money stored up from three lean years, during which there had been a freeze on departmental spending, that the Labour government was able to allocate this extra money without raising taxes.

Will I have enough to live on when I retire? Will I get proper health care when I am sick? Who will look after me when I am old? If I lose my job or become unable to work, will I end up in poverty or dependent on charity? These are significant questions for everyone, and across the world they are being asked with growing apprehension. For the burden on the state is only anticipated to get bigger. Growing elderly populations are already putting an ever-increasing burden on government (China will provide the most dramatic example of this after 2040), and global labor market pressures resulting from the technological revolution will only lead to more unemployed. And without adequate funding, will even governments genuinely committed to social justice and welfare be able to deliver what their societies need?

Attempts to raise money at the margins without advancing a sustained argument for higher taxes encourage voters to believe in a politics without difficult choices, and reduces room for change, especially now that the economic bubble has burst.

Ignoring one of the fundamental dilemmas of the capitalist age will not make it go away. International forces are undermining government's ability to sustain the welfare state, and its ability to restrain economic forces so that society could be a more humane and equitable place.

Abstaining from a discussion of the limits of Anglo-American-style capitalism, a system that favors the rich so blatantly and puts the profit motive above all else, is not a conscionable option. Suggesting an alternative way forward that apparently does not necessitate tough choices and is able to reconcile the goals of social justice and economic growth without asking the question of "economic growth for whom?" is fundamentally misleading.

Governments need to address the underlying issues and not evade them. But has the pursuit of economic success become so overriding a goal that any discussion of cost has become untenable? How far is society willing to go to gain a few extra points of economic growth? Where do our true priorities lie?

4

Manning the Door at the Private Sector HQ

Spies-R-Us

In 1947 British and American spy chiefs banded together to share security information. It was decided to operate a joint surveillance system, codename Echelon. Later three other English-speaking nations, Canada, Australia, and New Zealand, joined the project, although the USA remained the dominant partner.

The idea was sound: to renew the alliance that had successfully defeated Nazi Germany, in order to thwart a new threat, that of Soviet Russia. Moscow was on the verge of acquiring an enormous nuclear arsenal, and was trying as never before to spread her influence around the globe. The British and Americans concluded that they could better contain this new menace by working together. In America military Sigint (signals intelligence) units were established at Sugar Grove in West Virginia. In Britain a listening station at Menwith Hill, in Yorkshire in the north of England, far from London, became the most important international site for the group, and in particular for America's National Security Agency (NSA), the leading player in Echelon.

The history books have yet to detail just how significant a role the project played in countering the spread of Soviet communism. But no doubt it played a part, and in 1989 the Berlin Wall came down. By 1991 the whole Soviet bloc had crumbled, and the Communist threat had all but disappeared.

Echelon's electronic surveillance, however, did not; and a decade later, in February 2000, startling allegations emerged. Echelon was no longer being used for political and military espionage against dictatorships that threatened the free world. Instead it was being used to monitor everyday commercial activities of businesses belonging to some of America and Britain's closest allies.

Moreover, significant advances in technology during the 1990s meant that the system was now so powerful that it was reportedly capable of picking up every word of telephone, fax, and e-mail communications relayed by satellite anywhere in the world. Frighteningly, it applied to us all. Our each and every phone call and e-mail could be monitored. The implications were enormous. In a massive abuse of its original purpose, senior U.S., and possibly British, espionage chiefs used Echelon to spy on individuals and to pass on commercial secrets to American businesses.

These startling revelations came to light in February 2000, when newly declassified American Defense Department documents were posted on the Internet and for the first time provided official confirmation that such a global electronic eavesdropping operation existed at all. (The existence of Echelon had first been exposed in 1996 by a renegade agent in New Zealand but had not previously been proved.)

Within days the European Parliament released a report containing serious allegations. American corporations had, it was said, "stolen" contracts heading for European and Asian firms after the NSA intercepted conversations and data and then passed information to the U.S. Commerce Department for use by American firms. In Europe, the Airbus consortium and Thomson CSF of France were among the alleged losers. In Asia the United States used information gathered from its bases in Australia to win a half share of a significant Indonesian trade contract for AT&T that communication intercepts showed was initially going to NRC of Japan.

The European nations were furious, both with the Americans and with the British, their supposed partners in forging a new united Europe. In France a lawsuit was launched against the United States and Britain (on the grounds of breach of France's stringent privacy laws), in Italy and Denmark judicial and parliamentary investigations began, and in Germany members of the Bundestag demanded an inquiry. One Belgian member of the European Parliament summed up the feeling of many of his colleagues when he said that if the Americans and British had actually done this, it was "an intolerable attack on human rights." The Portuguese government—which at the time was chairing the European Union's rotating presidency—proposed establishing a European secretariat to monitor Echelon's activities.

The Europeans were stunned to discover that Big Brother was no longer Communist Russia or Red China, but Europe's supposed ally and partner, America, spying on European consumers and businesses for its own commercial gain.

The European Parliament's report stated that in 1995 the National Security Agency tapped calls between Thomson-CSF (now Thales Microsonics) and the Brazilian authorities relating to a lucrative $1.5 billion contract to create a satellite surveillance system for the Brazilian rainforest. The NSA gave details of Thomson's bid (and of the bribes the French had been offering to Brazilian officials) to an American rival, Raytheon Corporation, which later won the contract.

The report also disclosed that in 1993, the NSA intercepted calls between the European consortium Airbus, the national airline of Saudi Arabia, and the Saudi government. The contract, worth over $5 billion, later went to the American manufacturers Boeing and McDonnell Douglas.

Another target was the German wind generator manufacturer Enercon. In 1999 it developed what it thought was a secret invention enabling it to generate electricity from wind power at a far cheaper rate than had been achieved previously. However, when the company tried to market its invention in the United States, it was confronted by its American rival, Kenetech, which disclosed that it had already patented a virtually identical development. Kenetech subsequently filed a court order against Enercon banning the sale of its equipment in the United States. The allegations were confirmed by an anonymous NSA employee, who agreed to appear in silhouette on German television to reveal how he had stolen Enercon's secrets. He claimed that he had used satellite information to tap the telephone and computer modem lines that linked Enercon's research laboratory with its production unit. Detailed plans of the company's secret invention were then passed on to Kenetech.

German scientists at Mannheim University, who were reported to be developing a system enabling computer data to be stored on household adhesive tape instead of conventional CDs, began to resort to the cold war tactic of walking in the woods to discuss confidential subjects.

Security experts in Germany estimated that by the year 2000, American industrial espionage was costing German business annual losses of at least $10 billion through stolen inventions and development projects. Horst Teltschik, a senior BMW board member and a former security adviser to the former German Chancellor Helmut Kohl said, "We have discovered that industrial secrets are being siphoned off to an extent never experienced until now."

A Parisian lawyer, Jean-Pierre Millet, went further in the spring of 2000, claiming, "You can bet that every time a French government minister makes a mobile telephone call, it is recorded."

The orders, it seems, may have come from the very top. Early in his presidency, Bill Clinton defended the rights of business to engage in industrial espionage at an international level. "What is good for Boeing is good for America," he was quoted as saying.

Was this the new world order we had hoped for in the days of the cold war?

The New Politics

In fairness to the United States and Britain, virtually everyone was guilty of listening in on business. The European Union's report also revealed that France and Germany cooperate to eavesdrop on both North and South America, a claim backed up by insiders in Washington, who confirmed that European secret services pursue the same policy as the Americans.[1] China encourages its overseas students and scientists to pass commercial secrets back home; and the Japanese are known masters of commercial espionage.

In the post–cold war era, in the age of international laissez-faire capitalism, commercial and economic interests have tended to supersede all other national interests. Instead of acting as a check on corporations, governments are now doing all they can to romance them, acting less as night watchmen—the role Adam Smith said that they needed to play in order to ensure the success of free markets[2]—and more as round-the-clock doormen at the headquarters of Private Sector PLC.

Economics has become the new politics, and business is in the driving seat. Governments have redefined their role from that of rule maker to that of referee, from warden to corporate champion. Because of their dependence on the success of the private sector and exports for wealth, stability, rising aggregate standards of living, jobs—factors that can today be equated with political power—governments do not just sit back and let the market take its course. Instead they actively pursue policies that benefit business, giving up in the process their ability to set an independent agenda and favoring corporate Goliaths over individual Davids.

International politics in the twenty-first century is less and less about territorial gains, and more and more about increasing economic freedoms and market share.[3] In advanced economies governments now act as salesmen, promoting the fortunes of their own corporations in the hope of providing a core prosperity for their state[4] and keeping themselves in power.

To quote Madeleine Albright, former secretary of state: "Competition for the world's markets is fierce. Often, our firms go head-to-head with foreign competitors who are receiving active support from their own governments. A principal responsibility of the Department of State is to see that the interests of American companies and workers receive fair treatment, and that inequitable barriers to competition are overcome. Accordingly, the doors to the Department of State and our embassies around the world are open—and will remain open—to U.S. businesspeople seeking to share their ideas and to ask our help."[5]

The predominant concern of governments in the free trade, free market world of the early twenty-first century remains, even post–September 11, how to ensure their firms get a decent slice of the global economic pie. Witness the United States' blatant claiming of the spoils of the Iraq war by putting only U.S. companies on the initial shortlist of companies to reconstruct Iraq.

The Deal of the Century

In Russia, after the collapse of the Soviet Union, commercial interests quickly gained primacy. On September 20, 1994, the Russian state-owned oil company Lukoil signed an oil exploration contract with the Republic of Azerbaijan and a consortium of Western oil companies including BP, Agip, and Statoil, in what was billed as "the deal of the century," a contract that was expected to generate £100 billion in profits from the realization of Azerbaijan's vast oil reserves.[6]

What made this deal particularly unusual was the fact that Russia's involvement went entirely against the country's nationalistic sentiment. Since 1991 one of the pillars of official foreign oil policy of the Russian Federation had been its nonrecognition of the rights of other Caspian states to unilaterally exploit their offshore national resources.[7] Boris Yeltsin spoke on several occasions about the need to protect Russian strategic interests in the "near-abroad" of the old Soviet Union, and Pavel Grachev, the former Russian Defense minister, said on a visit to Turkey only a few months before the deal was signed that Turkey should keep out of "our" Azerbaijan.[8]

It seems, however, that even after the Russians had denounced the proposed deal as implicitly recognizing Azeri jurisdiction over the disputed waters, once they realized that they risked losing their share of the spoils, nationalistic foreign policy considerations were subordinated. At the very moment that the Russian ambassador and representatives of the Russian Ministry of Fuel and Energy were celebrating at the official deal signing ceremony in Baku,[9] the Russian Ministry of Foreign Affairs was declaring that Russia "will not recognize [the agreement], with all the ensuing consequences."[10] Empty words from a ministry that in the post–cold war era was clearly losing ground.

Not So Ethical Foreign Policy

Elsewhere in the world commercial interests are also being given primacy. Shortly after the electoral victory of Britain's Labour Party in 1997, Robin Cook, the new foreign secretary, approved a preexisting arms sale contract with the Republic of Indonesia. This was despite the official adoption of a government "ethical foreign policy," which prohibited the sale of arms to regimes that might use them for internal repression, the abuse of human rights or external aggression, and despite growing concern for massive human rights violations and fears of genocide in Indonesia's forcefully annexed territory of East Timor[11]—fears which later events proved justified.

According to Labour ministers the contract (for an estimated £438 million[12]), which included sixteen Hawk fighter jets manufactured by Britain's largest defense company, British Aerospace, would be "too hard to revoke."[13] This statement was somewhat misleading. The government could in fact have revoked licences without having had to pay damages or compensation to the parties involved. It was not prepared to do so because of the impact such a decision would have had on the British defense industry, an industry which it is estimated has sales of over £5 billion a year and employs over 400,000 people. Mr. Cook apparently had this clearly spelled out for him by both the prime minister's office[14] and BAE senior executives.

Nor was it only the British arms industry that stood to lose if the government blocked the arms sales to Indonesia. The U.K., Indonesia's second largest investor, hoped that the deal would enhance the interests of other sections of the British industry. At this time bidding was under way for billions of pounds' worth of contracts related to the giant Natuna gas field project, "one of the biggest investment opportunities in the world."[15] The prevailing wisdom seems to have been that Britain could not afford to risk its special relationship with Indonesia, whose imports from the U.K. had risen by 150 percent in less than five years.

Rather than discouraging sales to Indonesia for ethical reasons, the Labour government, since coming to power, has issued 125 export licences for arms to Indonesia for a wide range of products, including machine guns, military helmets, and aircraft spares. As late as July 1999, when British Aerospace Hawks were flying over Dili strafing the East Timorese, the U.K. had still not suspended the export of all military equipment to Indonesia. Such delays led to Britain being condemned by the East Timor Independence Movement as having been "the single worst obstructionist of any industrialized country."

To be fair to Robin Cook, it appears that he had earlier tried to change British policy. But in October 1998, when he tried to implement tough disclosure rules on arms exports, he was thwarted by then Industry Secretary Peter Mandelson, who put forward commercial objections, arguing that making licence applications public would identify companies' export plans and damage their competitiveness. More likely the true reason was to avoid political embarrassment for potential overseas customers or for Britain's massive arms export industry (the world's second largest after the United States).[16] As one of Robin Cook's aides commented, demands from the Department of Trade and Industry, which was bowing to "merciless" business pressures, had compromised Cook's ability to act.[17] Ethical considerations had little weight when pitted against a possible loss in sales and an erosion of market share.

The Indonesian case provides a clear example of the way that individual companies' interests and the aggregate interests of industry can now dictate foreign policy and override moral, humanitarian, and even legal considerations—Indonesia was violating international law in annexing and forcefully occupying East Timor. What is particularly noteworthy in this case is that the sales were not approved by a Conservative government but by a Labour one, by a party traditionally associated with a social conscience. In fact Tapol, the Indonesian human rights group, claims that the list of export licenses for Indonesia approved since Labour came into power bore a marked resemblance to the supply deals done under the Conservatives.[18]

Not only the Indonesians have benefited from Britain's trade policy. In Labour's first year in office numerous arms licences to ethically dubious regimes were approved: 84 to Pakistan, 42 to Sri Lanka, and 105 to Turkey.

The British public had come to expect this kind of behavior under previous Conservative governments. Conservative scandals had included Mrs. Thatcher's son Mark's alleged receipt of a multimillion-pound commission on the £20 billion Al Yamamah arms contract, which his mother signed with Saudi Arabia in 1985; and the £1.3 billion defense deal with Malaysia in 1988, later found to have been illegally and expensively linked with British aid. But Labour in opposition had always been a vocal critic of such deals. In fact Robin Cook, while he was shadow defense secretary, was the most vehement critic of the "Arms to Iraq" scandal in which a number of Conservative ministers were found to have sanctioned breaches in an embargo on the sale of arms to Iraq.

In the quest for market share the traditional lines separating left and right are disappearing. In the name of competitive markets, private interests are being served with little regard for ethics.

What About Human Rights?

It is not just that arms are being sold to repressive regimes; the whole idea of safeguarding human rights, a concept that was imbued with real meaning after World War II, has also fallen by the wayside as Western governments perceive only their need to promote trade and champion their firms' interests worldwide.[19]

A U.S. military contractor, DynCorp, accused of several human rights violations, has donated more than $150,000 to the Republican Party and was chosen by the Bush administration to police post-

Saddam Iraq. The European Union approved a customs union agreement with Turkey at the same time that the European Parliament was voicing concern on human rights violations and fears of genocide of the Kurds in Turkey's eastern territories. It did little for years about Nigeria's human rights violations under the regime of General Sani Abacha, apart from routinely condemning them. Trade and oil interests prevailed.

The United States backed the dictatorial Taliban regime in Afghanistan until 1997, despite its terrible record on human rights, its severe oppression of women, public executions, and intransigent Islamic fundamentalism. In large part, this was because the American oil company Unocal had signed a deal with the Taliban to build a $2 billion gas line and $2.5 billion oil line to transport oil and gas from Turkmenistan to Pakistan via Afghanistan. Washington eventually got tough with the Taliban—but only after immense pressure from the American feminist movement and because of the Taliban's support for the Saudi terrorist Osama bin Laden.[20]

Saudi Arabia's human rights record is almost as bad, but foreign governments unwilling to jeopardize their relationship with the oil-rich sheikdom continue to supply the country with equipment that can be used to torture or ill-treat prisoners. Between 1980 and 1993 the U.S. government authorized export licenses worth $5 million under the category OA82C, which includes thumb cuffs, leg irons, and shackles.[21]

The West continues to woo China despite its continuing poor record on human rights—its jailing of followers of the Falun Gong spiritual and exercise group, underground Christians, and dissidents; its ignoring of the internationally recognized rights of workers spelled out in the UN Convention on Human Rights; its use of forced prison labor—because of the huge market opportunity that China presents to Western firms. China has a good fifth of the world's population, and Hong Kong and Shanghai (with Singapore) lead economic and financial revitalization in the postcrisis East Asian economies[22] and are becoming the main "business hubs" of the region. It is not difficult to understand the attraction.

Bill Clinton, as a presidential candidate in the 1992 election, criticized then President George Bush for promising to renew China's most favored nation trading status, because of China's human rights abuses. But once he was in office he personally undertook the business of admitting China as a full partner in the WTO,[23] and by May 2000 he had succeeded in normalizing trade relations with China.

Clinton explained his about-face on China with the "trade encourages democracy" line, which claims that the greater the access to foreign business and influence the Chinese have, the quicker the Communist grip on the country will be loosened and the faster its attitude on human rights reformed—a view endorsed by various academic findings correlating investment by multinational firms with increasing levels of GDP, and correlating growing GDP over time with greater respect for human rights, freer markets, and even democratization.[24] Or at least that was the justification he used in 1994 when he took the radical step of uncoupling trade and human rights, so that he no longer considered the illiberal nature of the Chinese regime in his annual executive order extending most favored nation trading status to Beijing.

A correlation between foreign investment and democratization does not, of course, necessarily imply that foreign investment causes an improvement in local conditions. It is equally possible that foreign investment is attracted by an already improving political situation.[25] Furthermore, other academic studies dispute such conclusions, finding no meaningful relationship between levels of foreign investment and improvements in a country's human rights performance. One study carried out by the OECD[26] found that a country's desire to increase trade and investment could lead to a deterioration rather than an improvement in human rights, since, as we saw in the previous chapter, some governments feel that by failing to impose basic labor standards they can help attract inward investment.

Clearly the evidence is inconclusive. Yet it is not in question that countries such as the United States, which wield huge economic purchasing and investment potential, are in extremely strong bargaining positions. Surely if they were to make better human rights a condition of strong trade ties, developing countries would have to listen?[27] In the global economy could any country really afford to refuse these terms and turn their backs on trade with the West? Are Western governments not giving up the human rights issue unnecessarily, and failing to extract promises that they almost certainly could successfully demand?

If dangling the carrot of investment is considered an acceptable strategy, what about wielding the stick of sanctions? Shouldn't democratic governments dealing with repressive regimes at least consider the possibility of actively preventing trade and investment inflows?

Sanctions are gaining favor: Between 1945 and 1990 only two UN sanctions were imposed, on South Africa and Rhodesia (Zimbabwe), but over the past decade they have been used thirteen times. It remains unclear, however, whether they ultimately help or harm. Archbishop Desmond Tutu stated during South Africa's apartheid era:

> I have no hope of real change from this government unless they are forced. We face a catastrophe in this land, and only the action of the international community can save us . . . I call upon the international community to apply punitive sanctions against this government to help us establish a new South Africa—nonracial, democratic, participatory, and just. This is a nonviolent strategy to help us do so. . . .
>
> You hear so many extraordinary arguments. Sanctions don't work. Sanctions hurt those most of all you want to help. That is interesting. . . . I have to say that I find this new upsurge of altruism from those who suddenly discover they feel sorry for blacks

> very touching, though it's strange coming from those who have benefited from cheap black labor for many years. Spare us your crocodile tears, for your massive profits have been gained on the basis of black suffering and misery.[28]

The Burmese dissident Aung San Suu Kyi echoed Tutu's sentiments when she said in 1994: "If material betterment . . . is sought in ways that wound the human spirit, it can in the long run only lead to greater human suffering. The vast possibilities that a market economy can open up to developing countries can be realized only if economic reforms are undertaken within a framework that recognizes human needs . . . There are those who claim that the people of Burma are suffering as a consequence of sanctions, but that is not true. We want investments to be at the right time—when the benefits will go to the people of Burma, not just to a small, select elite connected to the government."[29]

Yet it is not clear that in all cases sanctions are the best option. Much criticism was attached to the use of sanctions against Saddam Hussein's Iraq, as was, for example, written in *The Guardian* in August 2000:

> The economy has been shattered and agricultural output badly disrupted. Malnutrition is endemic and medical services have been eroded. The UN's own agencies admit that up to 5,300 children are dying every month from disease, malnutrition, and related conditions. All told, it is believed that 500,000 Iraqis have died since 1991 as an indirect result of sanctions. Baghdad puts the figure at 1.5 million, or roughly 7.5 percent of the entire population.[30]

More understanding is needed before we can determine those cases where sanctions will be most likely to lead to the desired results, and those in which they will make the situation worse. Clearly the context in which they are imposed is important. What was appropriate in the case of South Africa, where it was essentially the white middle class who were most hurt by them, appears to have been wrong in the case of Iraq, where it was the poor who were most affected and where the sanctions arguably served to keep Saddam Hussein in power.

A framework is needed for determining the most apt response to a particular situation. But at present the only single unifying principle that seems to determine whether or not sanctions are imposed is whether or not they will harm the corporate interests of the country that is considering imposing them. How else could America's willingness to ban new investment in Burma by American companies, where U.S. business interests are virtually nonexistent, be reconciled with its failure to impose sanctions on China? Why wouldn't the "trade encourages democracy" line work there, too? Perhaps it is too much to expect consistency from a country that preaches human rights on the international stage yet ignores reports of rights violations in its own prisons, is the only advanced Western country that maintains the death penalty, and along with Saudi Arabia, Iran, Libya, and China is one of a handful of countries that still executes adolescents, the mentally ill, and the mentally retarded.

Cloaked in America's concern for human rights is a clear economic agenda. So the protests from human rights activists against Chinese policies toward Tibet and Taiwan that met China's Premier Zhu Rongji when he visited the United States[31] in 1999[32] fell on deaf ears in government. The dominant view was that "the bilateral relation with China is the most important that the United States now has,"[33] and that the United States "needs to improve relations with China and the best way to achieve this is to keep out of its domestic politics, in particular human rights and Taiwan."[34]

Cynics have long noted that American foreign policy is not driven by a concern for the greater good. U.S. government policy has long been dictated by corporate interests. But whereas during the cold war corporate interests masqueraded as military interests, this rationale is no longer convincing or relevant. In an ideal world a superpower like the United States, now that the Soviet military threat has disappeared, would concentrate on important issues like human rights. Instead, despite a growing rhetoric on human rights, we see little willingness to actively guarantee them. Will Iraq, for example, ever receive the level of aid it will need to liberate its people (not only from the tyranny of Saddam but also) from poverty and despair? This is unclear given that, at the time of this writing, more attention seems to be being paid to the dividing up of the lucrative postwar reconstruction contracts among such Bush-Administration-friendly U.S. companies as Haliburton, which Dick Cheney ran for five years; Bechtel, on whose board sits former Secretary of State George Schultz; and Fluor, whose VP, Kenneth Oscar, is a former army secretary and who oversaw the Pentagon's $35 billion procurement budget, than on ensuring the Iraqi people a viable economic future.

When governments evaluate trade, sanctions, and human rights on purely economic rather than ethical grounds and put their corporations' interests at center stage, they not only fail the people in the countries in question. They also fail to respect the wishes of many of their own citizens. Choice is restricted to what business or the markets want rather than the traditional democratic notion of what the people want.[35]

Confusing Democracy with Capitalism

But it is not just human rights that have been relegated; democracy has also lost out to trade interests for much of the past century. This is nowhere more striking than in the case of the United States of America, the world's loudest proponent of democracy, which has regularly allowed democracy to take a backseat to capitalism, despite its claims that it is its main priority.

When the tanks carried Yeltsin into Red Square in August 1991, George Bush said that democracy must prevail. What he probably didn't mean was a system in which people had the vote. It is unlikely that he really cared whether Russia was democratic; what he probably did care about was that a system should emerge in Russia that was favorable to U.S. interests and shared its economic values. A Singaporesque authoritarian system would in all likelihood have been just fine. Among potential investors in America and elsewhere in the West, the prevailing view was that "what Russia needs is a benevolent dictator"—that is, a dictator sympathetic to the American capitalist system.

America's official line during the cold war was: "The overreaching aim of our foreign policy is to spread democratic values." The truth, however, seems to be that foreign policy decisions were and are driven by a belief that the American system and its values are best protected by "a global system based on the needs of private capital, including the protection of private property and open access to markets."[36]

In Iran in 1953 the CIA backed the fall of the popular government of Prime Minister Mohammed Mosaddeq, who had been demanding that the Anglo-Iranian Oil company (the antecedent of BP) share more of its profits with Iran. Once the rule of Shah Mohammed Reza Pahlavi was restored, the returning ruler renegotiated his country's oil arrangements so that for the first time American oil companies were able to operate there, taking a 40 percent stake in the international consortium of private oil companies that were now to own and operate Iran's oil assets.[37]

In 1954 the United States helped overthrow the elected government of Guatelmala's President Jacobo Arbenz after he had expropriated 80 percent of the Tiquisate and Bananera plantations, then the American-owned United Fruit Company. U.S. State Department interest in the affair was intense, since the former law firm of the secretary of state, John Foster Dulles, represented United Fruit, and the head of the CIA, Allen Dulles, had been a member of the company's board of trustees.[38] Furthermore, Washington officials viewed this behavior as a serious threat to American investors' interests, and thus to American security. Arbenz's reformist government—which had undoubtedly made these expropriations in retaliation for the fact that none of United Fruit's profits had been reinvested or redistributed in Guatemala itself—was replaced by a CIA-backed military dictatorship in 1954. "Over the next forty years the military built the worst human rights record in the Western Hemisphere."[39]

In 1964 the United States encouraged the promilitary politicians José de Magalhães Pinto and Humberto de Alencar Castelo Branco in their successful attempt to overthrow Brazil's democratically elected government, a government whose espoused economic policies were again unacceptable to Washington, owing to its leftist leanings. "The new regime imposed military rule on Brazil for the next twenty years. During those two decades, the United States was the regime's best trading partner, while Brazil attracted more U.S. investment than any [other] Latin American country."[40]

Reagan took up the battle cry for democracy in the mid-1980s. The views of Jeanne Kirkpatrick, his ambassador to the UN, were central in reconciling the apparent paradox between actively supporting nondemocratic regimes, such as those of President Marcos of the Philippines, General Pinochet of Chile, and South Africa's pro-apartheid government, while continuing to demonize those in Cuba, the USSR, and China. According to Kirkpatrick there was no paradox once a distinction was made between authoritarianism and totalitarianism. Authoritarian regimes such as those in the Philippines, South Africa, and Chile were not democratic, often violently oppressed their peoples, and were usually corrupt but, because they shared American beliefs in open economic systems, it was acceptable for America to work with them. Totalitarian regimes, on the other hand, "were evil because they controlled every part of society, especially the economy, which was closed to private enterprise and foreign access."[41] If freedoms were ranked in order of priority, the first seemed to be freedom for American corporations to make money.

As we saw in the case of China, the Clinton administration continued to support the view (although not overtly) that the spread of capitalism was more important than the expansion of democracy. Again party politics seem to create few differences of opinion in a world in which economic interests are paramount. As Jacques Attali, former president of the European Bank for Reconstruction and Development, has put it: "The main mission of American diplomacy seems to be the 'export' of Western values including democracy—as long as doing so serves American interests."[42] Rather than democracy being inexorably linked to markets—as the Democrats suggested with the term they coined "democratic markets"—wherever the two ideals clashed, America favored the market over democracy.

Woodrow Wilson's proclamation that "the world must be made safe for democracy" has been presented as the driving ideology behind U.S. foreign policy for most of the last century. This is clearly misleading. When the American government talks about spreading democracy, what it really means is spreading its own flavor of liberal democracy. In fact its policies suggest that what it cares most about is just the "liberal" element or, even more narrowly, just the economic element of liberalism. It will encourage liberal attitudes to human rights only to the extent that they favor the development of a market economy—but in practice, as we have seen, it often prefers authoritarian regimes. Other elements of liberal democracy that it might value are the rule of law and the protection of private property, since investors need to feel secure from arbitrary seizures of their assets. But the "democratic" elements of liberal democracy—mass participation, an active civil society, regular elections—have proved much more expendable.

Throughout the last century, the United States has cloaked a foreign policy based on trade considerations, and centered on safeguarding private economic interests, in a veil of a concern for democracy. The leading U.S. diplomat in Asia in the early part of the twentieth century, Willard Straight, was probably closer to the truth when he observed that Americans make "politics out of money."[43]

WTO—Whose Trade Organization?

Nearly all capitalist countries do the same today. Competing economic interests have replaced ideological differences as the most divisive force in world politics. But whose interests are being fought over: corporations' or nations'? The answer today must be corporations, although nations consistently support them in fostering their interests overseas. Multinationals, many now as large and as powerful as many nation states, have a larger stake in the new world order than do many individual governments. And where the interests of corporations and states come into conflict, it is increasingly the corporate agenda that prevails.

Nowhere is this new imbalance of power between corporations and the state clearer than in an imposing 1930s building beside Lake Geneva. Here, at the headquarters of the World Trade Organization (WTO), rulings are made in the name of free markets that limit states' abilities to safeguard their people's interests, even in cases where states wish to do so.

In 1996, for example, Massachusetts resolved not to award public contracts to companies that do business with or in Burma, because of the country's appalling human rights record. Unilever, Siemens, and the Dutch banks ING and ABN-Amro were among a number of European companies that stood to lose out in consequence, and they lobbied Brussels in an attempt to thwart the decision. Thanks to their efforts, the European Union threatened to take the case to the WTO, arguing that the proposed ban was an unfair barrier to trade.[44] A British government spokeswoman said at the time: "This is not about Burma, it's about the United States and the application of international trade rules." Lawyers representing Massachusetts argued that had current trade rules been in force during the 1980s Nelson Mandela would still be in prison.[45]

Once it complies with the Association of South-East Asian Nations' (ASEAN) trade rules that will govern the Southeast Asian bloc from 2006, the Vietnamese government is unlikely to be able to maintain its ban on cigarette imports. Free trade policies would consider such a ban unfair, as indeed they did when China attempted to ban the importing of opium in the 1830s!

In 1996 the European Parliament voted 366 to 0 to ban synthetic hormones from beef, on the basis of strong evidence that they could cause cancer, reduce male fertility, and in some cases result in the premature onset of puberty in young children. Three months after the ban the U.S. government, responding to pressure from the agrochemical company Monsanto, the National Cattlemen's Association, the Dairy Export Council, the National Milk Producers' Federation, and other interest groups, complained to the WTO, arguing that the ban created a barrier to imports. In 1997 the WTO ruled in favor of the United States. The EU appealed but its appeal was rejected. In July 1999 the WTO authorized the United States and Canada to impose retaliatory trade sanctions worth over $125 million. European exports such as fruit juice, mustard, and Roquefort cheese were suddenly subject to massive tariffs.[46]

> In 1997, the European Union dropped its proposed bans on cosmetics which had been tested on animals and on fur from animals caught in leghole traps, for fear that the legislation would conflict with WTO rules. WTO guidelines will prevent it from banning imports of eggs from battery chickens, which means

> that U.S. producers will be able to undercut our farmers when battery cages are proscribed in Europe. The European Union . . . will probably have to drop its plans to make electronics manufacturers responsible for recycling their products.[47]

Time and time again, the WTO has intervened to prevent governments from using boycotts or punitive tariffs against companies that they have found to be acting in ethically unacceptable or environmentally unsound ways. In fact, in almost all environmental cases it has so far considered, the WTO has ruled in favor of corporate interests against the wishes of democratically elected governments.[48] Accountable to no one, it has restricted our choice over what we can eat, overridden laws passed by our democratically elected governments, started or sanctioned trade wars, and put our health at risk.[49]

Various aspects of the WTO have proved constructive; prior to its existence, trade disputes often led to prolonged, entrenched, and damaging economic warfare—in the 1930s, for example, the average world rate of tariffs was 45 percent, and wealthy countries had become protectionist and autarkic. Nevertheless, in imposing free trade doctrines upon states, the WTO makes it increasingly difficult for countries *not* to put trade interests first, even on occasions when their electorates or governments wish to give other interests primacy. National sovereignty is thereby weakened, not for a holistic greater global good but for a very particular one—that of American and European multinationals.

Behind Closed Doors

The WTO settles disputes in private. "When a challenge to a national or local law is brought before the WTO, the contending parties present their case in a secret hearing before a panel of three to five people who review written submissions and consider expert opinions."[50] The composition of the panel is determined by the Dispute Settlement Body, which almost always selects panel members and experts for their trade expertise. While many corporations advance their interests by participating in the expert meetings that are integral to negotiations, other input, such as testimonies, technical advice, and guidance on the environment or human rights, is purely discretionary and depends on whether a panel chooses to solicit them.[51] In contrast to the UN, environmental and other public interest groups are not allowed to observe WTO discussions, even when commercially valuable secrets are not at stake.

The burden is on the defendant to prove that the law in question is not a restriction of trade. Once a panel has decided that a domestic law does violate WTO rules, it may recommend that the offending country changes it. Countries that fail to make the change within a prescribed period face financial penalties, trade sanctions, or both. Panel reports become final in sixty days unless there is a unanimous consensus of the Dispute Settlement Body—all 144 member countries—to reject the report (highly unlikely), or the decision is appealed.

Developed and developing countries alike have been ruled against—both the EU and the United States have faced the WTO's censure—but developing countries often feel that they are treated as second-class citizens within the organization. At the Uruguay round of trade negotiations in 1993, the trade ministers of most third world countries were excluded from the final phase, despite the fact that the developing countries account for 80 percent of the WTO's membership. Stories of developing countries' trade ministers being "forced to wait for hours on end in the coffee bar, begging the emerging journalists to tell them the latest developments in the negotiations" later appeared in the international press.[52] In Seattle in 1999 African nations were understandably angered by the United States's decision to deprive their scheduled internal meeting of both translators and microphones. And before the WTO meeting in Doha in

November 2001, the Dominican Republic and Haitian ambassadors to the WTO were sent letters from the U.S. Trade Department saying that if their countries didn't sign up for the government procurement proposal, the Americans wouldn't look favorably on their aid packages.

Not only is the process unjust, but some thirty of the organization's 144 members cannot afford to base even a single representative at its Geneva headquarters. Switzerland promised in 1993 to finance a low-rent center to house representatives of developing countries, but the center has still not materialized. And developing countries cannot field negotiators in numbers to match those representing the developed world. While the European Union had 502 people in its delegation at Doha, for example, the Maldives had two, and St. Vincent only one. Third world nations are being forced to play a game in which the ante is unacceptably high and the openness of the forum is in question.

"I recommend that the system undergo some fundamental change," Panitchpakdi Supachia, Thailand's deputy prime minister, said in 1999. "It's high time we made it serve our development goals." Not surprisingly, Supachia's candidature for the top WTO job of director-general was initially opposed by the USA.[53]

While developing countries have far too little influence in the WTO, corporations have far too much, despite the fact that the WTO is ostensibly an organization of states. The very rules of the system have been established with corporate interests in mind and corporations themselves have played an increasingly significant part in shaping them. As James Enyart, a senior employee at Monsanto, has put it, "Industry has identified a major problem in international trade. It crafted a solution, reduced it to a concrete proposal, and sold it to our own and other governments. . . . The industries and traders of world commerce have simultaneously played the role of patients, diagnosticians, and the physicians."[54]

Corporations safeguard their interests by attending WTO ministerial conferences as members of national delegations. Preparations for the Seattle round were financed by soliciting large contributions from the private sector, in return for which business leaders were promised frequent access to the world leaders attending the conference.[55]

Helmet Maucher, the president of the International Chamber of Commerce (ICC)—an organization comprising seven thousand member-companies and representing the world's largest transnational corporations, including General Motors, Novartis, Bayer, and Nestlé—is pushing for his organization to be granted formal status within the WTO, despite the fact that at the moment only nation states are WTO members. "We want neither to be the secret girlfriend of the WTO," Maucher said in an interview, "nor should the ICC have to enter the World Trade Organization through the servants' entrance."[56] To pursue this more intimate relationship with the WTO, Maucher has made former GATT General Director Arthur Dunkel chairman of the ICC's commission on trade. Dunkel is also on the board of Nestlé.

But while the ICC waits for formal WTO status, the ability of big business and its representatives to sidestep the controls imposed by elected governments is already apparent, as evidenced by the growing dominance of free trade interests in international decision making.[57] In ancient Rome gladiators owned by the rich would fight to the death in the Colosseum. In Geneva nations more or less owned by corporations are pitted against each other in the WTO arena, unable to decide for themselves how they and their citizens would like to live and trade.

Chiquita La Bamba

Take the case of the U.S. multinational Chiquita. When the EU decided to award a quota of less than 10 percent of European banana imports to the company in order to protect small producers in former British and French Caribbean and African colonies, countries that rely on banana exports for survival, Chiquita persuaded U.S. trade representatives that the policy was unfair and harmful to the interests of the USA, claiming that the three largest American fruit multinationals—Chiquita, Dole, and Delmonte—would lose $520 million per year were this allowed to go ahead.[58]

Doubtless encouraged by the money donated by the corporation to both major political parties, the U.S. administration protested on Chiquita's behalf, lodging a formal complaint about the quota with the WTO and charging the EU with having a "discriminatory" approach to importing bananas. The U.S. government also threatened to impose a new 100 percent duty on a range of European products, from Walker's shortbread and Scottish cashmere sweaters to Italian light fittings, if the EU was not able to negotiate an agreeable settlement on banana imports. The negotiations, causing intense friction between the EU and United States, continued until June 1999 when the WTO ordered the EU to amend its quotas to allow more Central American bananas into Europe. The EU did not comply, and the United States instigated retaliatory sanctions in the form of punitive duties worth $191 million on such unrelated EU goods as bath salts, handbags, and bed linens.

Whether Carl Lindner, owner of the Chiquita fruit company, would have been able to put enough pressure on the U.S. administration to influence the international trade negotiations in the banana war case had he not been such a massive donor to the Republican and Democratic parties is debatable, especially as the Clinton administration's decision to pursue a trade war over bananas was sharply at odds with the way it had dealt with other comparable agricultural issues.[59] Lindner was undoubtedly a major contributor to party funds. Just before the United States had threatened to impose the tariffs on European imports, he had donated $200,000 to the Democratic National Committee. And only hours before the United

States lodged the formal complaint with the WTO, Chiquita had given $415,000 to state Democratic parties throughout the country. In fact it has been estimated that in all during the "Great Banana War," Lindner gave $4.2 million to Republicans and $1.4 million to Democrats.[60]

The acknowledged fact that in 1996 an entire village of six hundred people in Honduras, one of the Caribbean countries in which Chiquita operates, was bulldozed by government troops allegedly provided with food and fuel by Chiquita, seems to have been deemed irrelevant by the national and supranational decision makers; as was the fact that Chiquita had been involved in legal cases brought about by male workers who had allegedly been rendered sterile by frequent exposure to the pesticide DBCP (banned in the countries which produced it) on the plantations on which they worked[61]—more than twenty thousand cases of sterility have been reported. Objections to the United States' demands on the grounds that without access to European markets the economies of Caribbean islands would be devastated, increasing the economic pressures to produce drugs, and that for many Caribbean countries preferential access to EU markets is critical because small island producers cannot hope to compete on cost with the vast Central American plantations, were also set aside. Its protecting of the interests of Chiquita exemplifies the stance on trade that Washington has frequently taken toward Europe.

Puppets on a String

The Chiquita story exemplifies two major features of our new world order: the increasing degree to which governments and supposedly neutral global governance systems are dancing to the corporate tune; and the fact that global capitalism can often be a zero-sum game, with a gain by one side matched by a loss of another.

Once again we see that a world in which economic considerations are not balanced by other interests can be a grim one. In the global context the situation is particularly worrying. While in a national context, a finance ministry is typically only one ministry among others and decisions are reached collectively after different ministers have stated their case, international organizations such as the IMF and WTO, which shape and constrain the behavior of nation states, base their decisions purely on economic criteria, despite the fact that the implications of their decisions reach far beyond the domain of economics.

It is not that national statesmen and elected officials should not seek to bring prosperity and jobs to their own countries and act to ensure the robustness of their economies. It is that by giving corporate wishes such priority, by defining themselves solely in terms of economic success, by supporting international institutions that value economic interests above all else, governments are in danger of becoming the puppets of business. By defining political power in terms of economic power, democratic politicians lose sight of the reason they are elected—to serve all their constituents' needs, not solely the needs of big business. By determining foreign policy largely on the basis of commercial interests, governments forsake the opportunity to pursue other goals. Instead of creating a better world for people, governments work to create a better environment for business, under the mistaken belief that one will always lead to another, seemingly oblivious to the fact that in the age of globalization, multinational corporations lack national loyalties and cannot be relied upon to serve governments or national populations by delivering on either taxes or jobs. As a Colgate-Palmolive executive once explained, "The United States does not have an automatic call on our resources. There is no mind-set that puts this country first."[62] Or as Clive Allen, Nortel Network's executive vice president and chief legal officer, put it, "Just because we were born there [he was speaking about Canada] doesn't mean we'll remain there. Canadians shouldn't feel they own us. The place has to remain attractive for us to remain interested in staying there."[63]

But the needs of capital are not always the same as the needs of society. We, the people, risk being displaced—and in the one-ideology world of the twenty-first century, if things start going wrong, where can the global citizen go to be granted asylum? Is there a place where the sweeteners offered by the Chiquitas of this world are not accepted? Is there a place where these often conflicting priorities are better reconciled? Is there a place where governments can afford to remain impartial, or where politicians are not so dependent upon business? Can those of us living in the developed world at least draw comfort from the fact that, within the constraints imposed by globalization, democracy is working well enough for us?

5

Politics for Sale

Grandmother's Footsteps

Doris Haddock celebrated her ninetieth birthday on January 24, 2000, in Cumberland, Maryland. It was not a quiet affair. She had walked over three thousand miles across America to be there and celebrated by delivering a heartfelt speech on electoral reform to a large crowd. She then shared a birthday cake with local activists and, sent on her way by thirty local people singing "This Land Is Your Land," set off again for Washington, D.C. There, she gave three rousing speeches outside the Capitol and was arrested twice.

It is fair to say that Doris, known to her thousands of supporters as Granny D, is no ordinary nonagenarian.

When her husband, Jim, a lifelong grassroots political activist in Dublin, New Hampshire, died after sixty-two years of marriage, Granny D decided to overcome her grief by acting upon her and her husband's shared ideals in the most visible possible way. Five foot tall, with eleven grandchildren, she set off from Pasadena, California, on January 1, 1999, and walked the 3,200 miles to Washington to protest the growing corruption of America's political system by the vast donations made to the party machines by corporations and unions.

Supported financially and practically by the Fogies (Friends of Granny), she stayed with like-minded people, spoke to crowds and individuals, and gave interviews to the media as she went. Despite arthritis and emphysema she covered an average of ten miles a day, mostly on foot but, where there was snow, on skis.

In little over a year she trekked through California, Arizona, New Mexico, Texas, Arkansas, Tennessee, Kentucky, Ohio, West Virginia, Maryland, and Virginia to Washington, D.C. Early on, much of her journey was through empty desert, hundreds of miles from supporters and with scant media coverage, walking either alone or joined by one or two volunteers. "That never mattered," Granny D said at the time. "I had decided that I would go as a pilgrim."

But by the time she arrived at the Capitol in February 2000, over two thousand people were walking with her. She even made the pages of the political magazine *George,* appearing behind Hillary Clinton but ahead of Tipper Gore on its list of The Twenty Most Fascinating Women in Politics.

"It is, of course, a fool's errand," she said of her trek. "It is just an old woman walking across the land . . . talking about the kind of political reforms most people don't believe can really happen. But there are two things I would like you to understand about impossible missions. One is the fact that, sometimes, all you can do is put your body in front of a problem and stand there as witness to it. . . . The second is that there are no impossible causes on this earth if they are good causes. My dream of political reform will come true. I may live to see it from this side of life, or I will smile to see it from the other side."

Electoral campaign finance reform, pre the Enron debacle at least, was not a subject that typically garnered much support from within mainstream American politics, but at grassroots level Granny D's fear that politicians were being corrupted by the donations made to their campaign funds struck a very real chord. As she said at her birthday party in Cumberland, "A flood of special interest money has carried away our own representatives and our own senators, and all that is left of them—at least for those of us who do not write hundred-thousand-dollar checks—are the shadows of their cardboard cutouts."

On her website—www.grannyd.com—hundreds of people shared their experiences of joining her on her walk; the site still receives twenty thousand hits a day. Beth Kanter remembers "hanging out of the window at work, waving energetically and displaying homemade GO GRANNY GO! posters." John Parker caught up with her in Washington at the end of her trek: "To me, Granny D's speech was absolutely incredible. We had heard a number of other senators extol on how their brethren needed to change the laws. But when she took the microphone, she blistered everyone for having neglected their responsibilities. She must have spent 3,200 miles composing that speech, and she delivered it with some of the best oratory that I've ever heard. Despite her age, if she ran for office, I would vote for her."

Granny D's first arrest, in April 2000, was for attempting to read aloud the Declaration of Independence in the Rotunda of the Capitol. She was arrested again at the same place in July 2000 for trying, unsuccessfully, to read out the Bill of Rights. Manacled by the police, she was reportedly brought to tears when an officer tried unsuccessfully to wrest her wedding ring off her arthritic finger.

One of Granny D's main targets was so-called soft money—donations originally meant for incidental office expenses—which had become a major factor in American electioneering. Although direct campaign contributions had been banned in America since the Watergate scandal, a regulatory loophole meant that soft money was most commonly used for television advertising. In the 1992 campaign the two main parties raised $86 million in soft donations. By 1996 the figure had jumped to $260 million and analysts estimate that the 2000 election was financed to the tune of $393 million in soft money alone.[1]

Granny D's walk was initially aimed at supporting Wisconsin liberal Democrat Senator Russ Feingold, whose bill, coauthored by populist Republican John McCain, condemned the soft money system as "legalized bribery." The pair called for a voluntary ceiling on campaign finance, but in October 1999 they failed to gain enough Senate support to carry the bill forward to a vote. Granny D was, however, undaunted. In speeches she quotes the council of the Six Nations of the Iroquois: "Cast all self-interest into oblivion. . . . Your heart shall be filled with peace and good will and your mind filled with a yearning for the welfare of the people."

Unashamedly nostalgic, she has no qualms about wishing Teddy Roosevelt were still alive. Her campaign poster features her arm-in-arm with the former president: "I think I know what the great Republican would say to these dangerously overlarge monsters [large corporations], and I think I know what he would say about their trained monkeys in Congress," she said in Austin, Texas, in June 1999.

The court case following her Capitol demonstration sent out signals that at least parts of the American establishment may have been coming around to her point of view. In sentencing (a ten-dollar administrative charge and time served rather than the six-month sentence and $500 fine she had been facing), Chief Judge Hamilton of the D.C. federal district court said to her and her fellow demonstrators: "As you know, the strength of our great country lies in its Constitution and her laws and in her courts. But more fundamentally, the strength of our great country lies in the resolve of her citizens to stand up for what is right when the masses are silent. And, unfortunately, sometimes it becomes the lot of the few, sometimes like yourselves, to stand up for what's right when the masses are silent, because not always does the law move so fast and so judiciously as to always be right. But given the resolve of the citizens of this great country, in time, however slowly, the law will catch up eventually."

And so the law did. Thanks largely to Enron's collapse, which put a spotlight on political giving by corporate interests, the persistence of those lobbying for campaign-finance reform finally paid off. On March 20, 2002, the McCain-Feingold, Shays-Meehan Campaign Finance Bill passed the Senate and President Bush committed to sign it into law. Although it remains at the time of this writing unclear whether it will survive the constitutional challenge mounted against it, whether the Federal Election Commission will vigorously enforce the new rules, and whether in its weaker revised form the bill will have the hoped-for impact, Granny D was in the Senate gallery applauding when it was passed. "Was it worth it?" one of her fellow campaigners asked her before the applause ended. "Worth it? Yes, it was worth it," she replied.

The Price of Politics

But why do political parties in a democracy need to raise so much money in the first place? Because in the absence of clear ideological distinctions, parties can most effectively differentiate themselves in terms of marketing strategy and spending. Not only do politicians now defer to big business, politics itself emulates corporate tactics. Door-to-door canvassing, leafleting, and local meetings were the politics of yesteryear: low-cost, low-tech, and labor-intensive. The politics of today is expensive, businesslike, and capital-intensive, and it relies to a greater extent than ever before on mass communication via the media and advertising. Newspaper and magazine advertisements; terrestrial, cable, and satellite television commercials (where these are permitted); and Internet spots[2] are today's methods of reaching what we will see to be an ever more elusive electorate. Twenty-second ads are countered by twenty-second ads, and no party or politician can afford to be outspent by rivals.[3,4]

In the United States a phalanx of political consultants leads the way, seeking to engage voters with new and bolder ploys from direct mail campaigns to hard-hitting TV ads, with TV stations netting around $600 million in the 2000 election.[5] Remember the Bush ad campaign in which the word "rats" appeared subliminally in a broadcast targeting Democratic health care proposals? In Britain, by the 1990s advertising and public relations were already fully established as a part of political campaigning. "When the Conservative Party hired [ad company] Saatchi and Saatchi in 1978, it was headline news. By the end of the 1980s it would have been just as big news if a major party had chosen not to use professional marketing expertise in an election."[6] Political media advisers, advertisers, and image makers have become minor celebrities, their names probably as widely recognized as those of many cabinet ministers or congressmen.[7]

The cost of the new media circus is truly astronomical, especially in the United States. In the run-up to the 2000 presidential elections, the candidates seeking nomination raised and spent over $1 billion—the most in U.S. history—following the $651 million spent on campaigning in the 1996 congressional elections; and the 1998 House and Senate midterm elections, in which more than $1 billion was spent, seven times the total for the 1978 election, and almost double the 1992 amount. On average an individual Senate campaign now costs $6 million, meaning that each senator, as well as each defeated candidate, must raise an average of $2,750 every single day of his or her six-year term to pay for it.[8] Access to elected political office in the United States is now almost exclusively the privilege of the seriously rich. In his first four months of campaigning for the

2000 election, George W. Bush raised $37 million, more than either Bill Clinton or Bob Dole raised during the entire campaign in 1996. Jon Corzine, the former chairman of Goldman Sachs, spent $36 million of his own money to win a Senate seat;[9] Michael Bloomberg $50 million of his to become mayor of New York; and defeated candidate Michael Huffington laid out as much as $30 million when he stood for the Senate in California.[10] It is not possible to raise that kind of money from raffle tickets or barbecues. How can elections be free and fair when only the bankrolled can participate?

These developments are not unique to the United States. Although the figures are smaller, similar trends can be observed elsewhere. In an increasingly global political environment in which politicians are less able to deliver on actual policy and content, they need more money to capture their audience's attention. The 1997 general election was the most expensive British campaign to date, with the Labour Party spending just under £27 million and the Conservatives £28.5 million, double the amount spent in the 1992 election[11] (although under new legislation enacted under Blair, party campaign spending in an election year will now be capped at £20 million). In Taiwan the world's wealthiest political party, the Kuomintang (KMT), paid voters between fifteen and forty-five dollars to turn up at campaign rallies for the 2000 presidential election. In Russia the Our Home Is Russia party hired the German supermodel Claudia Schiffer and rap artist MC Hammer to provide support in the 1993 parliamentary elections.

Quid Pro Quo

This level of campaign spending is inherently problematic. The escalating costs of running campaigns and supporting political parties can no longer be met by membership contributions, union funds (where they are given), or personal donations. Even in countries that provide some degree of direct state funding to political parties, the funds provided by the state are nowhere near enough for today's political extravaganzas. "The democratic political process costs money—in ever increasing amounts."[12] So who do politicians turn to, to meet the shortfall? As Grandma would say, the private sector, of course.

All over the world from Moscow to Paris, from Washington to London, corporations and businesspeople are bankrolling politicians and political parties. Parties and candidates are given support; money is contributed to campaigns; political rhetoric is publicly endorsed; unwritten IOUs are registered.

Such funding comes from a small elite. In the United States, for example, "only one quarter of 1 percent of the population gave two hundred dollars or more to congressional candidates or the political parties in the 1995–'96 election cycle and 96 percent of the American people [didn't] give a dime to any politician or party at the federal level."[13] America's largest five hundred corporations, on the other hand, gave over $260 million to the Democrats and Republicans from 1987 through 1996.

Of course, corporations are not in the business of giving something for nothing. Money buys action and influence.[14] In exchange for amounts of money that are often quite small from their point of view, they expect a significant return. As Supreme Court Justice David Souter has said, "There is certainly an appearance . . . that large contributors are simply going to get a better service, whatever that service may be, from a politician than the average contributor, let alone no contributor."[15] So when Charles Keating, the boss of an American thrift company, Lincoln Savings and Loan, that later defaulted and cost the American government and taxpayer hundreds of billions of dollars,[16] was asked whether the $1.3 million he had donated to five senators' campaigns had influenced their behavior, he replied, "I certainly hope so."[17] Mr. Keating could afford to be frank, because his contributions were above board and entirely legal.

Despite their legality, such donations undoubtedly create an undesirable opaqueness. It is always difficult to prove a link between corporate funding and policy changes that favor a donor company, but the string of unproven connections and unlikely coincidences that link financial contributions and favorable policy changes is becoming just too long to explain away. Chiquita was no anomaly.

In the United States, where the problem is probably more widespread and of a greater magnitude than elsewhere in the developed Western world, we see countless examples of probable cause and effect. For example, the Center for Public Integrity draws attention to the 1994 Fair Trade in Financial Services Act, which had been lobbied for by NationsBank, for whom the legislation would mean savings of $50 million a year. "Two weeks later, the cash-strapped Democratic National Committee (DNC) received a $3.5 million line of credit from the NationsBank at a favorable interest rate."[18]

There are many similar examples. In April 2003 two U.S. senators, Larry Haig and John Breaux, endorsed a letter from the U.S. Sugar Association to the president of the World Health Organization (WHO), Gro Harlem Bruntland. It threatened that unless the WHO scrapped its about-to-be-published guidelines on healthy eating, which said that sugar should account for no more than 10 percent of a healthy diet, they would demand that Congress end its funding of the WHO. Haig is the Senate's number two recipient of campaign contributions from the Sugar Cane and Sugar Beets industry. One of Breaux's most significant campaign contributors in 1998 was Joseph E. Seagram and Sons—a multimedia, sports and recreation, and beverage company.

And many other prominent American politicians have been tainted by allegations that they have given preferential access to their corporate backers. In 1992, for example, House Democratic Leader Richard Gephardt of Missouri persuaded President Clinton not to tax beer as a means of financing his proposed health care plan; since 1988 Gephardt has received over $300,000 in campaign contributions from the Anheuser-Busch Company, the country's largest brewer.[19] Even Bill Bradley and John McCain, who both stood on campaign finance reform tickets in the presidential primaries, seem tainted. Mr. Bradley, while senator for New Jersey, apparently supported forty-five special bills aimed at offering tariff reductions and export aid to companies producing highly toxic chemicals. During that same period, chemical firms were among the biggest donors to his election fund.[20] Senator McCain was accused of intervening with the Federal Communications Commission on behalf of a major contributor to his campaign, Paxson Communications, and of trying to stop the expansion of a national park in Nevada—a move which would have benefited the property development company Del Webb, his seventh biggest sponsor.[21]

The protection given to tobacco interests in the United States, although recently undermined by product liability litigation in some states, further illustrates the influence that money can buy. "From 1987 through 1996, the tobacco companies have contributed more than $30 million in contributions to members of Congress and the two major political parties. In 1997, a single sentence added to a mammoth tax bill by Republican House and Senate leaders Newt Gingrich and Trent Lott, gave a $50 billion tax credit to the tobacco industry."

The list of links between campaign donations and votes in Congress is almost endless. Jennifer Shecter of the Center for Responsive Politics collates campaign contributions and the resultant votes by legislators who receive them. She notes: "The ten House and ten Senate members who received the largest contributions from the American sugar industry all voted to preserve a sugar quota that keeps prices high for consumers. Similar matchups are made for the timber industry, the B-2 bomber, the gambling industry, and even drunk-driving legislation, among others."[22] In fact it has been argued that the very reason why Microsoft's monopoly was ever addressed by the U.S. government was that Bill Gates did not join the campaign and lobbying bandwagon soon enough—Microsoft had, as recently as 1995, no Washington office.[23] A strategy that was subsequently reversed, with Microsoft over the three years of investigations and litigation nearly tripling its campaign contributions and more than doubling its lobbying expenditures.[24] And it was to good effect: the Justice Department decided to no longer seek a breakup of the computer giant—a decision very much in line with the pledge made by Bush in his attempts to woo Silicon Valley prior to the 2000 elections when he promised to support "innovation over litigation every time."

George W is, of course, the king of the "revolving doors" school of politics, having recruited key officials to his administration direct from the nation's boardrooms. Dick Cheney was headhunted from the oil services company Haliburton. Karl Rove, Bush's chief political strategist, had been chief political strategist for Philip Morris from 1991 to 1996; Mitchell Daniels, the head of the White House Office of Management and Budget, is a former vice president of Eli Lilly; and the treasury secretary, Paul O'Neill, came from the giant aluminum manufacturer Alcoa.

And since taking up office Bush has passed a series of laws that appear to favor big business. He has scrapped a raft of work safety measures which had been negotiated between the federal government and the unions for much of the previous decade. He proposed a bankruptcy bill, long demanded by the banks and credit card companies who sponsored Bush and his party to the tune of over $25 million, whose effect will be to strip Americans who have declared themselves bankrupt from some of the legal protection they have from their financial creditors. And he has passed a number of measures intent on protecting the interests of the energy companies that bankrolled his campaign. In addition to withdrawing from the Kyoto Protocol on global warming, the president reversed several executive directives passed in the final days of the Clinton administration, which aimed to protect 58 million acres of federal land by restricting logging and road building; reneged on his own campaign pledge to regulate carbon dioxide emissions from power plants; discussed the opening up of the vast and virgin Arctic wilderness in Alaska for prospecting and drilling; cancelled a looming deadline for automakers to develop prototypes for high-mileage cars; rolled back safeguards for storing nuclear waste; proposed shifting Superfund hazardous-waste cleanup costs from polluters to taxpayers; blocked a program to stem the discharge of raw sewage into America's waters; undermined protections for national parks and national monuments; and reversed Clinton decrees on clean-air standards for buses and big trucks. And that's just a partial list.

He also allowed Enron executives (good judgment here—not) to vet candidates for the commission regulating the U.S. energy markets, filling the vacant Republican seats on the commission with commissioners who had the backing of Enron and other power companies.[25] The interests of the American people were suborned to those of the major U.S. energy giants that bankrolled him: $25.4 million was all it cost.[26]

Politics has been on sale even in issues with potential national security implications. In 1997, for example, President Clinton overrode the objections of the Justice Department and permitted an American company, Loral Space and Communications, to export technology to China that would allow it to improve its nuclear missile capabilities, granting a waiver to sanctions that had been imposed after the 1989 massacre of hundreds of prodemocracy protesters in Tiananmen Square.[27] Bernard Schwartz, chairman of Loral, was the largest personal donor to the Democrats that year. President Clinton, of course, denied any quid pro quo. "The decisions we made were made because we thought they were in the interests of the American people," he later said. Which American people was that, exactly?

And it has been argued that the Bush administration initially blocked Secret Service investigations into Islamic terrorism because of the influence of powerful oil corporations, many of whom had stumped up wads of money for the Bush campaign. John O'Neill, former head of the FBI's counterterrorism office in New York, who later became head of security at the World Trade Center and was killed in the September 11 attacks, left his FBI job complaining that his investigations into Al Qaeda had been obstructed, stating that "the main obstacles to investigating Islamic terrorism were U.S. corporate oil interests and the role played by Saudi Arabia."[28]

And Across the Ocean . . .

In the United States, where the level of funds needed to finance political campaigns is particularly excessive, however distasteful or dangerous the outcome may be, at least all sides seem to be operating within the confines of the law. In many other countries, however, money has been shown to turn the wheels of policy making in contexts whose legality is frequently in question.

In fact in almost all advanced democracies in recent years, major political parties and senior statesmen have been implicated in covert deals, many of which are related to issues of access for large corporations. In Belgium, for example, a scandal was uncovered in 1993 involving the payment of more than $3 million in bribes by the Italian helicopter firm Agusta and the French aerospace manufacturer Dassault in exchange for the securing of orders for equipment from the Belgian armed forces.[29] Senior politicians in Spain have been convicted of channeling money into secret funds and their own bank accounts. France has been plagued by a series of substantial illegal payments by companies to all the leading domestic political parties, which led to an outright ban on corporate donations in 1993 and a reform of party funding rules.[30] In 1999 the French Finance Minister Dominique Strauss-Kahn resigned, following moves by examining magistrates to investigate a payment of about $90,000 from a left-wing students' health insurance fund.[31]

Revelations in Germany surrounding former Chancellor Helmut Kohl have exposed the venality of a system that formerly had a reputation for cleanliness. A major part of the corruption scandals engulfing Germany's Christian Democrats revolve around Helmut Kohl's alleged offers to companies in exchange for significant donations. A parliamentary committee is currently investigating claims that the Kohl government took millions in bribes from the French oil giant Elf Aquitane during the early 1990s, in payment for the company's takeover of the East German Luena oil refinery and a chain of gas stations in the former Communist east. The German chancellery has disclosed that all documents relating to the debacle have gone missing. In Italy the entire political system has come under suspicion; in the early 1990s 40 percent of members of Parliament were under investigation for corruption.[32] It is hardly surprising that only 19 percent of Italians say they are fairly or very satisfied with their country's democratic performance in the 1990s.[33] The ability of business to use donations as a way of buying influence or securing lucrative contracts is also substantial in Japan, where in 1991, despite the reform of campaign finance rules, a steel frame maker was found to have given Y80 million ($600,000) to a former minister in return for government contracts.[34] Endemic corruption brought about the end the thirty-eight-year rule of the Japanese Liberal Democratic Party in 1993.

In Britain the Conservative government was brought down at least in part by its association with financial "sleaze," prompting a government inquiry into the funding of political parties. In 1989, for example, John Major, then Britain's chancellor of the exchequer, refused to even consider a bill that would have required foreigners doing business in Britain to pay taxes on their earnings there. There may have been no connection, but it is undoubtedly true that a number of those who would have been hit were significant donors to the Tory party.[35] By 1994 a Gallup poll found that nearly two thirds of the British public believed that "most MPs make a lot of money by using public office improperly."[36]

That belief seemed verified by subsequent events. Before the 1997 general election, a number of Tory MPs were found guilty of receiving cash for asking parliamentary questions from lobbyist Ian Greer, who represented private business interests. Businessman and owner of the prestigious Harrods store, Mohamed Al Fayed recalls hiring Greer after being told by him at their first meeting, "You can rent them [members of Parliament], you can do what you want, you can rent them exactly like taxi drivers. They will do anything for you and I can handle this for you."[37] Further revelations from the Neil Hamilton–Al Fayed libel trial revealed the truth of Greer's claim. Hamilton, a former Tory minister, was found to have accepted envelopes stuffed with fifty-pound notes, Christmas hampers, Harrods gift vouchers, and a family holiday at the Ritz Hotel in Paris in exchange for asking specific parliamentary questions, signing early-day motions, and meeting ministers, all on Al Fayed's behalf.

The new Labour government did not escape the taint of sleaze. In 1998, soon after its election victory, controversy erupted over the issue of a political donation. Before the election, Labour had received £1 million from Bernie Ecclestone, the man behind Formula One motor racing. Once in office, the party dropped its stated opposition to tobacco sponsorship of motor racing. When the story came out, the potential link between the donation and the policy shift clearly made the payment suspect. Peter Mandelson, the Northern Ireland secretary, was forced to resign in the "cash for passports" debacle of January 2001 when it was suggested that he had inappropriately intervened in the application for British citizenship by S. P. Hinduja, who had given £1 million to New Labour's ill-fated Millennium Dome. And in February 2002 Tony Blair faced controversy over his links to the Indian steel magnate Lakshmi Mittal, a Labour party donor, who had allegedly won a bid in Romania thanks to British government support.

The connection between corporate donations and policy decisions is, of course, not always clear-cut. While the American gun lobby is a large contributor to political campaigns—in 2000 it hosted the biggest fundraiser ever in American politics, raising $21.3 million for the Republicans in one night—it also has genuine mass support. And many politicians support local industries not only because they donate money to campaigns, but because jobs, and therefore votes, are tied up in these industries. "Dairy state representatives like Senator Pat Leahy, a Vermont Democrat, will vote to protect dairy interests whether or not they receive cash contributions from dairymen."[38]

Whether or not definite causal links can be found, what is clear is that by making money available to politicians, or bankrolling the armies of lobbyists that now fill the corridors of power, corporations and businesspeople are at least ensuring that politicians listen to their demands and consider what they want. After all, why would business bother if they didn't think donations were likely to serve their interests generally, if not specifically through bills or policy decisions? But by permitting a system that allows such transfers to take place, politicians are essentially admitting their inability to control "corporate creep"—their willingness to rate the interests of certain groups higher than those of others.

Read It in the Papers

Even if comprehensive limits on campaign spending and the financial involvement of corporate interests were introduced, as is happening in the U.K. and to some degree at least in the United States, political playing fields would probably never be entirely level. For it is not just money that politicians need from business. Given a diminishing interest in politics, an increasing distrust of politicians, and the limited loyalty people now feel for particular parties (discussed later in this chapter), politicians more than ever need the endorsement of influential external forces to either gain or retain the electorate's support. And no external factor has more influence than the media. A favorable editorial, or an item on the evening news, allows a politician to communicate with thousands or millions of people, for free, via a medium that many voters are likely to trust more than paid-for political advertising. And the more limits that are put on financial support, the greater the extent to which media influence becomes important, and the greater the incentive for parties to seek other means of support of a nonfinancial kind.

In Britain it is the print media that wields immense political power. The extent to which the owner of the *Sun* and *The Times,* media baron Rupert Murdoch, whose News Corporation owns 30 percent of British newspaper circulation[39] and whose papers span the social spectrum, has been wooed by first Conservative and then Labour leaders in Britain, as well as by politicians in other countries where he has media interests, attests to his enormous influence.

In fact Tony Blair's wooing of Murdoch before the 1997 general election has been seen by many as a crucial element in Labour's winning strategy. Rumor has it that in 1994 Alastair Campbell would take the offered job of press secretary to new party leader Tony Blair only if Blair would look through some "documents." The documents were front pages of the *Sun* from around the times of the previous two (lost) elections, all of which endorsed the Conservatives, including the one with the now famous headline "It Was the *Sun* Wot Won It!" published the day after Neil Kinnock's 1992 defeat. Campbell's message was clear—if we don't get Murdoch on our side, we won't win.

Blair agreed, Campbell came onboard, and New Labour's campaign to win over the newspaper magnate began. In 1995 Blair was invited by Murdoch to address his annual News International conference at Hayman Island, a resort off Australia's Queensland coast. No deal was struck at the conference, although it is believed that Murdoch was told that his empire would be safe under New Labour.[40] Six weeks before the 1997 election, the *Sun* made a historic shift away from the Conservative Party, telling its 4 million readers: "The *Sun* Backs Blair,"[41] assuring them, a week before the election, that "there is no doubting his conviction." Blair's courting of Murdoch seemed to have paid off.

Although Blair denies paying a price to Murdoch for the *Sun*'s election support, the latter was no doubt delighted by the Labour leader's confirmation that he would not impose new restrictions on cross-media ownership,[42] as well as by the Blair government's refusal to back a House of Lords initiative to introduce legislation to curb "predatory pricing" of newspapers, the practice of taking a loss on the cover price which enabled *The Times* to double its circulation over a five-year period to the detriment of its competitors.

Arguably, the Conservatives had repaid Murdoch rather more substantially for his previously staunch support. Murdoch's 1981 acquisition of *The Times* and *Sunday Times,* for example, was never referred to the Monopolies Commission, despite the fact that it gave him the control of four national newspapers.

For many years before the Murdoch–Blair rapport, Murdoch had been, in the words of Charles Douglas-Hume, a former editor of *The Times,* one of "the main powers behind the Thatcher throne."[43] "Rupert and Mrs. Thatcher consult regularly on every important matter of policy, especially as they relate to his economic and political interests. Around here he's often jokingly referred to as 'Mr. Prime Minister,' except that it's no longer all that much of a joke. In many respects he *is* the phantom prime minister of the country."

Somewhat ironic, then, that by 1997 Murdoch was supporting Blair's New Labour. But given Murdoch's response when asked whether he intended to live up to the promises he had made in connection with his battle for control of *The Times,* his switch of sides should not perhaps be too much of a surprise: "One thing you must understand. You tell these bloody politicians whatever they want to hear, and once the deal is done you don't worry about it. They're not going to chase after you later if they suddenly decide what you said was not what they wanted to hear. Otherwise they're made to look bad, and they can't abide that. So they just stick their heads up their arses and wait for the blow to pass."[44]

Which might help explain the fact that by May 1, 2000, the front page of the *Sun* under the headline MAYDAY, MAYDAY was warning its readers that the Blair administration was beginning to lose the next election. Blair had not fulfilled expectations, the newspaper claimed, the people were feeling let down. The Conservatives' William Hague was a real threat.

Blair, on vacation at the time, sent the newspaper a 975-word handwritten defense. The three-page letter made the front page the following day, under the headline RATTLED.

In the United States we see much less blatant wooing of media moguls by politicians—the U.S. media environment with its lack of national newspapers, a much less partisan press, and the freedom for political parties to advertise on TV means that candidates are less dependent on unpaid advertising than they are in the U.K. But American media companies, in exchange for their funding of political campaigns, have, like any other corporate donor, received protection for their bottom line. That includes protection of lucrative tobacco ads in newspapers and magazines, the government giveaway of up to $70 billion worth of broadcast spectrum space, a way of dodging free airtime for political candidates,[45] and the easing of caps on television station ownership.[46] The media lobbies in the States enjoy considerable success.

A Hundred Years on . . .

In the twenty-first-century world of global capitalism, while nations compete for investment flows and the jobs and growth that corporations can provide and politicians need ever greater funds to compete with their rivals to win over the electorate, governments actively do what they can to promote the interests of business. They court corporations, sponsor their causes, pander to their needs. Rather than keeping them in check, governments are providing them with countless alibis. Rather than aiming to control and limit their activities, governments are allowing business to help shape them and their policies.

In 1876 presidential candidate Rutherford B. Hayes remarked of his government, "It is a government of corporations, by corporations, and for corporations."[47] In the chaotic and blatantly corrupt environment of late-nineteenth-century America, large companies were virtually able to buy legislation. As Matthew Josephson put it in his classic study of early American capitalism, *The Robber Barons,* "The halls of legislation were transformed into a mart where the price of votes was haggled over, and laws made to order were bought and sold."[48]

Over a hundred years later the situation seems broadly similar, not only in the United States, which has a well-charted history of corruption and pork-barrel politics, but elsewhere, too. As we enter the new millennium, arguably the entire world is *of* (international) corporations, *by* (international) corporations, and *for* (international) corporations. The problem is the same, its geographical extent significantly worse. Corporations have, in effect, begun to lay down with force what is and what is not permissible for politicians all over the world to do.

The End of Politics

With governments, regardless of their political persuasion, increasingly impotent, unwilling, or unable to intervene on their citizens' behalf, and seemingly having lost any sense of moral purpose, it is hardly surprising that the electorate is turning its back on conventional politics, even in countries that proclaim democracy as one of their greatest achievements. No wonder that their citizens are ignoring the ballot box, and parliament itself, as a means of registering their demands and protests. People are growing more distant from political parties and more critical of political institutions.[49] Never since the development of the mass franchise has there been such disengagement from politics.

The soundbite culture in which politics now operates has debased political rhetoric and helps to make voters feel that political institutions are increasingly irrelevant to their lives. The issues discussed in parliaments rarely have much to do with their concerns. In the U.K., Gallup polling since 1991 has consistently shown that people see the most urgent problems facing Britain as health, education, the cost of living, and unemployment.[50] Yet debates in the House of Commons are dominated by controversies over the European Union—of little interest to most voters—or legalistic debates that seem to fail to engage the interest of many MPs, let alone the general public.[51]

Politicians are increasingly seen as impotent, irrelevant, and dishonest. People see their governments as unable to deliver what they promise, obsessed with unimportant issues and internal politicking, riddled with corruption, clinging to outmoded notions of authority, and increasingly in the pockets of businesspeople. The distinction between incompetence and dishonesty is becoming blurred as, in country after country, senior politicians are discovered to have engaged in corrupt practices.

In today's "one-ideology" world, where parties have such similar policies on key issues such as taxation and welfare that it is almost impossible to differentiate clearly between them, voters are failing to develop an enduring sense of party identity and are increasingly unwilling to offer long-term loyalty to any political party. If current trends continue, less than half the American population will identify with *any* political party, making election results increasingly volatile and unpredictable.

Many people have simply lost faith in politics. In the United

States a 1997 poll showed only 14 percent rating the honesty and ethical standards of congressmen as "high" or "very high"—only narrowly beating car salesmen and advertising executives;[52] and by the beginning of the new millennium roughly three in four Americans didn't trust the government to do what is right most of the time.[53] In the U.K. the percentage of the electorate who had "great confidence" or "quite a lot of confidence" in parliament dropped from 54 percent to 10 percent between 1983 and 1996.[54] And 60 percent of French respondents polled have expressed "no confidence" in political parties.

All over the world, from the developed democracies of the United States and Western Europe to Latin America and the Far Eastern countries, poll respondents have less confidence in their institutions of government than they had a decade earlier.[55] In all advanced industrial democracies the public is effectively detaching itself from party loyalty and disengaging from politics. Most people, wherever they live, seem to believe that elected officials don't care about their concerns.[56]

Boycotting Politics

Alienated, dissatisfied, and skeptical voters are boycotting politics, even in countries where democracy has a long history. In the United States only 51 percent of voters turned out in the 2000 presidential election, despite the knowledge that it would be a close run; 55 percent had voted in 1992.[57] And only 39 percent of eighteen- to thirty-four-year-old American college graduates in 2000 said they had given a lot of thought to that year's election, down from 68 percent in June 1992, while turnout at the November 2002 midterm elections of 39 percent was the lowest voter turnout in the established democratic world for elections of a national legislature.[58] The public's sense of resigned alienation also manifests itself in other ways, such as tuning out of public affairs altogether. For example, 40 percent of the American people did not know the name of the vice president of the United States.[59] And during the 2000 party conventions season in the United States about the same number of people entered the Republican and Democrat websites as clicked onto budweiser.com.

The elections for the European Parliament in 1999 saw less than 50 percent of the EU's 297 million electorate bothering to vote, down from 57 percent in 1994. In the U.K. the turnout was only 24 percent of registered voters[60]—in the same week that one million British people bothered to vote to change the Choco Krispies brand name back to Coco Pops. And the "landslide" victory for Labour in the U.K. general election of 2001 was achieved on a turnout of 59 percent of the voting-age population, down ten points from 1997, sixteen from 1992, and the lowest turnout since World War II. Less people voted for any of the British political parties than voted in the final round of the U.K.'s version of *Big Brother.*

Even in Eastern European countries that only finally managed to become democracies in 1989–'91, turnout has been falling. In Poland turnout started out at 64 percent in 1989 but had fallen to 49 percent by 1997. In the Czech Republic turnout in 1990 was 93 percent, but it has fallen at every election since then and stood at 77 percent in 1998. In Hungary turnout fell from a high of 76 percent in 1990 to 60 percent in the 1998 election.

Membership of political parties in the United States, Germany,[61] France,[62] in fact pretty much anywhere in the developed world, is lower than at any time since the war. In 1950s Britain, for example, the Labour Party claimed a membership of about a million; today that is down to 360,000. Over the same period membership of the Conservative Party declined from some 2.8 million to less than half a million.[63] By comparison, the Royal Society for the Protection of Birds currently has over a million members.

How are the politicians responding? Desperate to woo those who will vote, and to reassert their legitimacy, politicians and political parties are turning to business to provide them with the tools. Pollsters, image consultants, and advertising specialists, such as James Carville, Stanley Greenberg, and Philip Gould—the team who helped Clinton to victory in 1992 and have subsequently been active in Britain, Germany, and Israel—are being brought in to advise politicians on how to project themselves to increasingly less interested electorates. "From Latin America, through Europe to India and Australia . . . professional media consultants have swelled in number and increased in influence. Where once electoral strategy was determined by party leaders it is now increasingly influenced by the media professionals, relying on in-depth focus group surveys . . . direct mail, and market research."

What influence do politicians really have in a world of global capitalism? Single-handedly, not much, as we have seen. Governments are now like flies caught in the intricate web of the market. And voters see their powerlessness. They sense that politicians' hands are tied and that their promises are increasingly empty. They watch politicians dancing to corporations' tunes. They are aware that the political rhetoric they hear is not being translated into any sort of actual reality; they feel that in many cases politicians have entered into a covert pact with business. And so, increasingly, they are turning their backs on politics.

As the political class clamors ever more loudly to be heard and goes to ever greater financial lengths to be noticed, evidence suggests that voters have stopped listening. In this ideologically singular world, where democratic parties and politicians are becoming increasingly homogeneous, and in which the people's interests are being usurped by those of business, the people are registering their discontent by not voting. After the long fight for universal franchise, the great-granddaughters of women who chained themselves to railings for the vote are now making their political statement by refusing to buy politics.

6

Shop, Don't Vote

Eco-Shoppers Save Butterflies

In the autumn of 1999 there was little doubt about which topic environmental pressure group Greenpeace should choose for its annual conference. Fears about genetically modified (GM) food crops, dubbed "Frankenstein foods" by the media, had dominated British headlines throughout the summer, and footage of eco-warrior activists with scythes and billhooks destroying fields of GM crops had become a familiar sight on news bulletins.

Scheduled to appear were, in the green corner, Lord Melchett, organic farmer and former Labour minister; and against him Robert Shapiro, the chairman and CEO of Monsanto, the world's leading producer of GM seed.

Lord Melchett, executive director of Greenpeace, was at the time facing criminal charges for tearing up a test crop of GM maize on a farm in Norfolk in July that summer. But where once Shapiro would have appeared in person at the conference and evangelized about the cleaner, healthier future offered by GM, he changed his mind at the last minute and instead spoke by satellite link from America. Onlookers found him grim, defensive, and defeated. He as good as apologized to Lord Melchett and Greenpeace's 2.5 million members. "Our confidence in this technology and our enthusiasm for it has, I think, been seen widely, and understandably so, as condescension and indeed arrogance," he said. "Because we thought it was our job to persuade, too often we forgot to listen."

After a bruising year of protests, campaigns, and consumer revolt, Shapiro was facing up to the fact that Europe, and particularly Britain, were rejecting Monsanto's biotechnology. And things were to get much worse for the American company, which had invested $8 billion in its attempt to dominate global GM biotechnology. Its mission statement, to help people "lead longer, healthier lives, at costs that they and their nations can afford and without continued environmental degradation," had done little to convince a British public once-bitten by the BSE scandal and deeply mistrustful of further scientific "improvements" to agriculture. Furthermore, opposition to GM had spread rapidly through Europe and beyond, to Japan, Mexico, and Brazil, finally provoking outcry back in the United States.

Shapiro was right to be downbeat. The company's share price had fallen steadily from $50.88 in May to a low of $36.48 in December 1999, at which stage the drug manufacturers Pharmacia and Upjohn announced a merger with Monsanto, attracted not by its GM interests but by its highly profitable pharmaceutical arm. Agribusiness, once the jewel in Monsanto's crown, was spun off into a much smaller stand-alone company accountable to its own shareholders. Robert Shapiro was to remain as Monsanto's CEO only until the merger was complete.

Monsanto's GM soybeans, potatoes, and cotton were first

launched in the United States in 1995 and had provoked only the mildest of protests from America's environmentalists. Despite the premium cost, sales increased rapidly and within three years 25 percent of corn, 38 percent of soybeans, and 45 percent of cotton grown in the United States was GM, almost all from Monsanto seeds, altered to make them resistant to its bestselling herbicide, Roundup.

In Britain, however, the response was very different. The European Commission approved European imports of GM products in 1996 but withheld permission for GM seeds to be planted commercially. Even this was enough to provoke alarm. At first the opposition was sporadic, with letters to *The Times* in 1997 from sources as diverse as the National Union of Farmers and the National Council of Women.

Monsanto then made perhaps its biggest mistake, launching a heavy-handed £1 million advertising campaign which only served to focus and fuel opposition. The GM issue moved from the letters pages to the front pages, and was soon dominating British news bulletins and headlines. Protests against the ill-advised campaign were upheld by the Advertising Standards Authority, which criticized the company for misleading the public and presenting opinions as fact.

Middle-class England was, for once, wholly in agreement with the eco-warriors. In April 1999 the *Daily Mail* launched "Gene Food Watch," a major campaign highlighting the use of GM ingredients—mostly, at this stage, processed food "tainted" by American imports of Monsanto's soy and corn. Consumer fears translated into falling sales, and that spring Marks & Spencer became the first U.K. main street retailer to remove all GM foods from its shelves. By the end of May 1999 Tesco, Sainsbury's, Safeway, and Asda had followed suit, as had most of the country's major food producers. The chairman of Birds Eye Walls, Iain Ferguson, said, "We have taken this decision in direct response to the wishes of a growing number of consumers in the U.K."

In the summer of 1999 America's (and the world's) largest miller,

Archer Daniels Midland, bowed to European consumer pressure and instructed its farmers to separate GM from non-GM export crops. British alarm began to filter back to America, especially after a huge public outcry in May 1999. Researchers at Cornell University had revealed that pollen from GM corn (soon known as "killer corn") could prove deadly to the endangered monarch butterfly. Monsanto was suddenly under attack from all sides.

Back in Europe, Deutsche Bank issued a stark warning to would-be investors in GM companies. "Genetically modified organisms have crossed the line. Today the term GM becomes a liability," said its analysts in August 1999. As a final indignity, in October that year the Food and Drug Administration reacted to the outcry by announcing plans to hold public hearings into the long-term effects of GM products.

In November 1999 Shapiro admitted publicly that Monsanto had "hopelessly underestimated" European consumer fears. The company then tried hard to promote its "skylark-friendly" sugar beet in an attempt to convince environmentalists; but it was a case of too little, too late.

In February 2000 Tony Blair finally backtracked, having only a year earlier dismissed the public's concern with the claim that "there [was] no scientific evidence on which to justify a ban on GM foods and crops," admitting there was indeed "potential for harm" to health and the environment. He announced a three-year moratorium on commercial planting of GM crops, hailed by campaigners as a U-turn. Scientific tests of GM plants, however, continued to be given the green light, with twenty-five new trial sites announced in August 2000. In the long term, it may prove more difficult to resist the spread of GM foods than protesters hope, but that is scant comfort for Robert Shapiro and Monsanto.

Supermarket Activism

In the world of the Silent Takeover, many citizens of democratic societies feel that their governments are no longer looking out for them, so many of them are increasingly looking out for themselves. If the state is perceived as no longer to be relied upon to ensure the quality and safety of the food we eat, the air we breathe, or other environment issues, a growing number of people are beginning to bypass traditional political channels and express concerns and demands directly to the bodies that are believed to be able to address their concerns, the corporations.

Cases like that of Monsanto have taught us that, just as we once possessed political power by the combined impact of our votes, today we can wield real power over corporations by combining with other consumers. The outcome of the British public's reaction to GM foods—the near collapse of Monsanto, and the banning of GM products by most food manufacturers and retailers in the U.K.—happened not because politicians willed it, but because consumers, aided by the media, did.

The Monsanto case is not unique. On both sides of the Atlantic, people are banding together and using the threat of boycott and negative publicity not only to pressure corporations into changing the way that they do business, but also to address the flaws of the system itself. If governments, pleading the constraints of globalization and the need to pursue economic growth, and taking corporate PR at face value, are perceived as failing to monitor how corporations are making money and to put limits on their activities, people increasingly will. In Roswell, a suburban community of Atlanta, Georgia, local residents have been protesting against advertising billboards. "It's a visual pollution," says Jay Litton, a resident of six years. "They are destroying Roswell." The residents have launched a successful campaign against the companies who advertise on and own the billboards. "These companies are making money out of us by polluting the air. . . . Those companies are making millions of dollars in profit from advertising here. Everyone is profiting but the taxpayer."[1]

Litton's words echo the feelings of a growing number of people who are beginning to ask who exactly is profiting from the capitalist dream. The increased indifference to party politics seems not to be matched by a growing lack of interest in a wider world. A lack of faith in traditional politics should not be confused with apathy or disengagement from society. Thatcher was wrong when she said there was no such thing as society; society has in fact proved remarkably durable. During the same period that trust in political authority has been fading, membership of grassroots movements has been on the rise—in the United States the percentage of people involved in community action in 2000 was three times that in the early 1990s[2]—a rise due to individuals' inability to gain recognition in the public arena by conventional means, and to a loss of faith in politicians' ability to champion their interests or make any difference to their lot. People no longer believe that politicians can resist the force of nonelected organizations. They have lost faith in politicians' ability to put the people's interests first.

Believing that their governments are weakening in the face of multinational hegemony, seeing that the nation state is no longer the focus of power in the world, sensing that politicians are no longer leading business but that business is telling the politicians what they can and cannot do, we no longer concentrate our efforts on politicians.[3,4] Instead, more of us are going straight to the new political power—business.

Increasingly the most effective way to be political is not to register one's demands and wants at the ballot box, where one's vote depends on the process of representation, but to do so at the supermarket, where a dollar spent or withheld can, cumulatively, lead to the desired end, or vocally at a shareholders' meeting. These forms of direct action are replacing rather than complementing conventional forms of political expression. All over the developed, democratic world, people are shopping rather than voting.

I Am What I Buy

The use of consumer pressure to keep corporations in check is not new. For as long as there has been a market in commodities, there have been champions of the interests of those who consume. From the Cooperative Movement in nineteenth-century Britain, via the American Progressives at the turn of the century, to Ralph Nader and the National Consumer Council, producer interests have been called to account for the price exacted for their goods, whether in terms of money or in their effects on safety, health, or human rights.

In the past, however, it was usually only a small minority who engaged in consumer boycotts and political shopping and then only on isolated occasions. Today there is a sense that consumer activism is beginning to enter the mainstream, that because of the increased visibility of corporate actions and the corresponding invisibility of political will, and because of recent major popular successes, there is an increasing realization that tarnishing the corporate image of unethical companies, or leaving their products on the shelves, are powerful weapons. In 2000, 25 percent of Americans surveyed said that they would be willing to join a boycott for or against a particular cause—an increase of 50 percent from the early nineties.[5] In an era of political apathy and disengagement, consumerism is beginning to replace citizenship as the tool through which ordinary people gain identity and recognition in the public arena: "We took decisions, which in retrospect were mistakes. We now realize that alone we could never have hoped to reach the right approach—that we should have discussed them in a more open and frank way with others in order to reach acceptable solutions. . . . In essence, we were somewhat slow in understanding that environmentalist groups, consumer groups, and so on were tending to acquire authority. Meanwhile those groups we were used to dealing with [e.g., government and industry organizations] were tending to lose authority."

So said Shell's President C. A. J. Herkstroter in late 1995, after the Brent Spar oil storage structure, which the company had planned to dump at sea, had instead, as a result of public pressure, been moved to a Norwegian fjord for dismantling.[6]

The year 1995 was seminal for consumer activism, and Shell was the company under the spotlight. Setting a pattern that would later be mirrored in most of the later consumer victories, the media played an active role. In this case Greenpeace, a central player in the new consumerism, masterfully engineered a media campaign to protest against Shell's decision to sink the redundant Brent Spar in the deep Atlantic. The Greenpeace-media alliance was extremely effective, buoying up public resistance. German consumers boycotted Shell gas stations—on one day in the summer of 1995 sales fell by 50 percent—creating a serious concern for senior management and damaging the company's image. Britain's Conservative government, as might be expected, backed the company, assuring the public that Shell's solution was the least environmentally damaging one. In this instance, their claim proved to be true, but too late to prevent the damage to Shell.

It was not only over environmental issues that consumers were beginning to act without reference to their elected politicians. While politics was becoming further removed from ethical issues, shopping was becoming imbued with a sense of morality. The new cathedrals of the middle classes are shopping malls with parking for thousands. And the new religion practiced by a growing number of these shoppers is consumerism with an ethical slant, a stance actually endorsed in the U.K. by the church. The Church of England–approved prayer book for the millennium, *New Start Worship,* counsels the faithful that "where we shop, how we shop, and what we buy is a living statement of what we believe. . . . Shopping which involves the shopper in making ethical and religious judgments may be nearer to the worship God requires than any number of pious prayers in church. . . . If we take our roles as God's stewards seriously, shoppers collectively are a very powerful group . . ."

The liturgy then goes on to say: "If, when we ourselves are not on the poverty line, we always go for the cheapest price, without considering that this price is achieved through ethically unacceptable working conditions somewhere in the world, we are making a statement about our understanding of the word 'neighbour.' "[7]

The consideration of how a product is made soon became the next manifestation of enlightened consumerism. In April 1996 the American public was shaken to discover that Wal-Mart's Kathie Lee collection of clothing—endorsed by their favorite talk show host Kathie Lee Gifford, who was paid $5 million a year to do so—was being stitched by teams of Honduran children, some as young as thirteen, working twenty-hour days for wages as low as thirty-one cents an hour.

Gifford burst into tears on national television when confronted with the evidence, and she quickly became an active campaigner against sweatshops. Over the next few months she set up a charity; talked Wal-Mart into changing its policies; personally went to the factories to ensure that they were no longer hiring children; gave three hundred dollars to each of the rather bewildered factory workers at Seo Fashions in New York's Chinatown, who it emerged had also been making blouses for her line under sweatshop conditions; and testified before Congress at a subcommittee hearing on child labor. The conference on sweatshops the following day was attended by supermodel Cheryl Tiegs, the heads of Levi Strauss, Wal-Mart, Kmart, and dozens of other corporate chains who were there largely in response to Kathie Lee's public plea that they "do the right thing" as well as the then labor secretary Robert Reich, who said, "We will look back to today, years from now, and say that this was a major turning point in our collective commitment to rid the nation—and also even the world—of sweatshops."

Kathie Lee Gifford in tears on national TV was a turning point. Corporations began to realize that they could no longer use their distant suppliers as scapegoats: Western consumers were demanding that similar rights be extended to overseas workers as to those at home. But on the whole, the corporate response was reactive. Following the Wal-Mart saga came exposure of the practices of Nike, Gap, and Disney, among others. Each case provoked similar consumer actions and only then damage control reactions from the firms. Five years on, the issue still makes the front pages, but this time the target is university-endorsed clothing and sportswear, not mass-market brand names, and the protesters are not A-list celebrities but eighteen-to-twenty-two-year-olds. Protests, described as the biggest act of student unrest since the anti-Vietnam campaign of the 1960s, are taking place on university campuses throughout the United States with students objecting to factory conditions in Guatemala, Nicaragua, and Bangladesh where, they claim, exploitation is commonplace in the garment industry. In the first six months of 2000 alone, college students led sixteen sit-ins at university buildings in protest. And their demands are being heard. The Universities of Oregon and Michigan and Brown University all withdrew their membership of the industry-backed Fair Labor Association, which the students claimed to be ineffective, and instead signed up with the Workers' Rights Consortium, a body made up of students, university officials, and labor and human rights campaigners—but not government officials—that will monitor how the clothes are made.

The globalization of production over the past years has been matched by an increasing globalization of information and concern, although a clear information asymmetry exists between corporation and consumer, with consumers unable to react until a story breaks. But each time a company *is* exposed, we see images of six-year-old children bowed over workbenches, or adults crammed thirty at a time into squalid dormitories in the brief intervals from stitching the sneakers, footballs, and sweatshirts that we wear and play with.

These images in themselves seem to create an upsurge of a sense of responsibility. Molly McGrath from the University of Wisconsin sweatshop campaign has said, "A lot of the students involved in the campaign are generally like me, white kids who've never been involved in political or social justice activity before, but who recognize that our clothes shouldn't be made by people who are treated like slaves—it's an easy thing to understand." Easy because, unlike traditional political concerns such as health care or defense which are complex, multidimensional, and require sophisticated analyses of multiple trade-offs, a single issue like this is easy to grasp.

While politicians are allowing corporations increasingly free rein, and while traditional voting is seen to be increasingly ineffective as a means of political expression, shopping has been imbued with a new political significance. It is the most effective weapon in the armory of ordinary citizens, enabling people to press for some degree of accountability in governments, international organizations, and multinational corporations. In the world of the Silent Takeover, in which the social contract between government and the people is increasingly meaningless, popular pressure is doing something that governments can't or won't: demanding that corporations be judged by noneconomic criteria, holding them accountable in a way that we cannot hold our elected representatives.

The New Consumerism

Consumer politics offers us a chance as individuals to make a difference, a chance to exercise direct power in a way denied to us by contemporary representative democracy.

But how significant is this trend? Are a growing proportion of us political shoppers paying the premium for Fairtrade coffee and "child-labor-free" footballs? Or are we merely paying lip service to good causes and willing to revolt only from the comfort of our own armchairs? Is serious activism, as it was in the 1960s, seventies, and eighties, still a minority cause?

Sunday morning. Central London, home. I wake up to the excesses of the night before. Dirty dishes in piles. I open my bottle of Ecover and squeeze biodegradable liquid onto yesterday's plates crusted with residues of GM-free organic pizza. Fill a mug with Fairtrade coffee and boil a free-range egg. Take a "not tested on animals" Lush bubble bath. Pull on my "child-labor-free" Reeboks, "made by 100 percent union labor," Levi's, and "never use furs" Chloe T-shirt. Spray my hair with a Wella non-CFC canister. Read the papers and learn about the latest McDonald's boycott. Remind myself to pick up a leaflet from the protesters on my next outing by jotting down a note on my pad of recycled paper.

Nip down to the Body Shop to get my "fairly traded" moisturizer, read in-store leaflet on globalization while paying for it with my "investments only in ethical companies" Co-Operative Bank credit card. Stop to fill up with unleaded gas on my way home. Two gas stations on either side of the road. Same prices, same gas. Remember that the one on the left has been involved in an oil spill in Nigeria. Turn right with no second thought. Come home and log on. Check mail on "we put social issues first" AOL. Send off a standard form e-mail to McDonald's, protesting their activities in Argentina. Enter the UN hunger site,[8] click my mouse, and silently thank American Express for donating that day's bowl of rice and mealies. All the while snacking on Ben & Jerry's "we don't cut down trees in the Amazon" ice cream.

And I am not alone. A U.S. survey in 1995 revealed that over 75 percent of Americans would boycott stores selling goods produced in sweatshops. Almost 85 percent said they would be willing to pay up to a dollar more on a twenty-dollar garment if it carried a label guaranteeing that it had been made under humane conditions.[9] A Gallup poll in Britain the same year found that three out of five consumers are prepared to boycott stores or products because they are concerned about their ethical standards, or have already done so.[10]

More recent polls have confirmed these earlier findings. A study carried out in the States in 1999[11] showed that between 40 and 50 million Americans, about 25 percent of the adult population, are beginning to make such value-based choices in more and more product categories. A Mintel poll in the U.K., also in 1999, found that three quarters of respondents would make their choice of products on a green or ethical basis.[12] And a poll carried out by the Henley Center in 2002, which measured the percentage of Britons who said they had chosen—or boycotted—a product or company for ethical reasons in the last twelve months found that 29 percent of respondents said that they had done so. When price and quality are comparable, socially responsible businesses have the advantage.

A survey carried out at the end of 1999 of twenty-five thousand people in twenty-three countries confirmed these findings. Consumers worldwide are more likely to base their impressions of a company on its labor practices and ethical conduct than on the quality of its product or service or its finances. Asked to say in their own words what formed their impressions, those surveyed put issues related to corporate social responsibility at the top of their lists. And 20 percent of consumers worldwide—40 percent in North America—said they had acted on their perceptions. They told pollsters they had avoided the products or services of companies they saw as socially irresponsible, or had spoken negatively about those companies.

Once the preserve of left-wing activists who drank low-grade coffee because it was cooperatively produced, ethical and political values are now being factored into the buying decisions of a significant proportion of the population. In a world in which people no longer feel that their elected representatives will do the right things on their behalf, they are increasingly asking this of corporations; and it appears that, when confronted, corporations are prepared to deliver.

The business world is much more responsive than the political one. Consumers in the private sphere constantly make decisions about how and where to buy, and businesses—at least successful ones—have learned to react swiftly to their consumers' desires.

But why have we seen such an increase in consumer activism since 1995? Why are we seeing a burgeoning of activity on the part of consumers now?

Timing undoubtedly plays a part in the explanation. By the mid-1990s the Western middle classes were in financial terms doing better than ever before. In the developed world, at least, their material needs were well-sated and they felt physically safe now that the threat of Soviet attack had been laid to rest. This "feel-good factor" lessened their need to focus on themselves, and it allowed for a shift in emphasis toward other issues such as quality of life and care for others—distant strangers and future generations, too, not only people in their own neighborhood: concerns that seemingly no longer troubled the state.

Another factor was the strategic decision taken by pressure groups and nongovernmental organizations to shift away from government-focused campaigns and coopt the media in an attempt to mold public sentiment and thus force accountability from corporations.

Shrewd pressure groups have realized that global companies can be used as a lever to influence the actions of their host countries. Campaigners against the regime in Burma in 1997, for example, saw that while their own governments were unresponsive, they were able to have a significant impact by identifying corporations with major operations there and threatening a worldwide consumer boycott.[13] When in the summer of 1995 Australians sought to protest against French nuclear testing in the South Pacific, environmental groups, in contrast to earlier campaigns, didn't ask their government to get involved. They called for a consumer boycott of French products. A national free telephone service was run by Consumer Power, a group formed solely to campaign on this issue, offering advice to consumers on what products to avoid.

This new consumerism focuses on the corporations themselves, rather than using governments as a conduit. Human Rights Watch published in 1999 a damning report not on a government, but on a corporation, Enron, and its activities in India.[14] And increasingly activists are redirecting resources away from lobbying and government-oriented campaigns and using the mass media and advertising campaigns to appeal directly to consumers. By generating negative publicity against target corporations, they can create public support for boycotts or other forms of consumer protest.

Lord Melchett highlights the need for business-focused campaigning: "A lot of international regulation arises out of what business will deliver—it's not politicians leading business, it's business telling the politicians what they can do. If you're campaigning for change, you have to get business to change, and then the politicians will follow."[15]

The best way of getting business to change is by getting consumers onboard, which is a much easier prospect than ever before. Satellite, digital, cable, and the five-hundred-channel universe have shortened the distance between the child laborer in Saipan, the domestic retailer, and the individual shopper. Images of seals being clubbed to death for the sake of a fur coat, misery inside factories in faraway places, and computer simulations of "Frankenfoods" graphically depict to consumers the result of inactivity.

Each successful campaign encourages others. People are more likely to demand companies to reform the way they are doing business because of the success of similar campaigns. The students protesting on American campuses have seen previous similar protests deliver results. And people have learned that corporations, unlike politicians, nowadays respond quickly to their demands and seem to take them seriously. It took twenty years of consumer pressure for Barclays Bank to pull out of apartheid South Africa, but

Shell moved the Brent Spar platform within months, and Marks & Spencer removed GM foods from its shelves overnight.

Making use of the Internet, the speed with which consumers can now communicate with each other or with pressure groups and with which corporations can get things done, makes a mockery of the time taken by governments to effect any changes in policy. Unlike politicians, corporations cannot afford not to keep their customer base happy, let alone make them actively unhappy. They don't have the luxury of guaranteed stays in office.

Shareholder Activism

It is not only consumers who are steering corporations in new directions. A growing number of individual investors and, more importantly, financial institutions, are choosing to use their power as shareholders to "regulate" corporate maneuvers.

Over $1 trillion is currently invested in the United States in managed portfolios that use at least one social investment strategy, a thirty-fold increase since 1984. One dollar of every eight invested in the United States is now invested in ethical funds. Ethical investments by the British public have exploded from almost nothing in 1980 to more than £2.8 billion today, with new launches of ethical funds almost monthly. Projections are that they will triple again within the next five years.[16] In Australia ethical investment is the fastest-growing sector in the managed fund market, with major companies joining the trend. In Germany the influx of money into ethical and environmental funds has risen thirty-six-fold over the past two years.[17]

While ethical funds vary from the palest to the deepest of green, most avoid investing in companies involved in tobacco, arms, nuclear power, and animal experiments, and many will reject companies which are active in or linked to oppressive regimes that do not respect human rights. All these, of course, are issues that our governments have failed to address adequately.

How ironic that *Homo economicus,* who was only ever supposed to be interested in maximizing his own self-interest, has turned out to be so interested in investing in the common good.

And how depressing that politicians, who were expected to focus on realizing the public's needs, in the world of the Silent Takeover have so often remained mute.

Individuals' investments in ethical funds are still small compared to the total investment funds in Britain, accounting for only about 1 percent of the total U.K. market, so their impact is small. But pension funds wielding huge amounts of money look likely to have an increasing impact on how companies do business. More than 50 percent of American investment funds are managed by pension funds; the figure is between 70 and 80 percent in the U.K. Investments made by these institutions are so large that they have to view them in the long term. Research findings over the past few years have positively correlated social and environmental performance with financial performance over time, so that a corporation's commitment to these values is increasingly being factored into the fund manager's purchasing decisions.[18] Not so politicians who, as we saw in the cases of the U.K. government and Monsanto and Brent Spar, were still regarding as top-dollar stocks those that the market and any reputable fund manager had already heavily discounted.

Despite its reluctance to champion social and environmental matters too openly, Britain's New Labour government introduced legislation in July 2000 whereby British pension fund trustees now have to report on whether they take account of social and environmental matters. Although it is not mandatory for them to adopt green or ethical criteria, they will have to justify their decisions if they do not. Several of the U.K.'s leading pension funds, in light of this new act, stated a clear commitment to factoring in social and ethical issues when considering which companies to invest in. "We are taking this approach for financial reasons, not moral ones. We believe that environmentally sound companies are better run and will make good investments in the long run," said a spokesman from Prudential Portfolio Managers, which manages £150 billion worldwide.[19] By changing the payoff structure, has the British government acted by stealth to curb the behavior of corporations?

Institutional investors now wield power over companies, not only by boycotting them if their activities do not please, but in subtler and more powerful ways. In the same way that corporations are "buying" stakes in political parties and "hiring" politicians to champion their causes, shareholder activists—whether pension funds, environmental groups, pressure groups, or activist funds—are using their power as shareholders to buy holdings in companies with weak ethical records specifically so that they can effect real change from within.

In the United States the Interfaith Center on Corporate Responsibility is a shareholder advocacy organization that counts among its members more than 275 institutional investors of the Jewish, Catholic, and Protestant communities, whose combined assets are in excess of $110 billion. Members buy shares in target companies and then use shareholder resolutions to force votes on social issues at shareholder meetings. Hot issues include promoting equal employment practices, dogging the tobacco industry, and challenging U.S. corporations to improve working conditions in factories outside the country. In this way, seventy-eight-year-old Sister Patricia Marshall of the Sisters of the Blessed Sacrament managed to get Anheuser-Busch to drop several of its stereotypical images of Native Americans from its advertising for Miller beer in the United States. Other successes have included getting PepsiCo to sell its bottling plant in Burma; Kimberly-Clark, makers of Kleenex, to sell off their tobacco operations; and 3M, America's third-largest billboard company, to phase out tobacco advertising.

Not only nuns and rabbis are standing up to protest in shareholder meetings. Other activist pension funds, investment funds, and environmental groups are also adopting this strategy. "Corporations often ignore stakeholders, but they listen to shareholders," said Brent Blackwelder, president of Friends of the Earth. "By aligning ourselves with investors . . . we can greatly increase our chances of being heard and heeded by belligerent companies."

Often, the purpose of, for example, turning up at shareholder meetings dressed as polar bears—as a coalition of environmentalists, investment groups, and pension funds did at BP's annual meeting in April 2000, to protest against the company's exploration activities in the Arctic Ocean—is not to get a resolution passed, but to generate publicity and so shame and pressure the company into action. In this case the motion gained the support of only 13 percent of voting shareholders; but since BP was just about to relaunch its company brand as environmentally minded and careful of natural resources, the protest couldn't have come at a worse time. As a survey of Canada's top one hundred CEOs revealed in 1996, "Shareholder activism and unpredictable questions top the list of 'worst nightmares' at annual meetings." It is a "nightmare" that seems to be ongoing. In April 2003 a group of British environmental and human rights groups launched a concerted three-month campaign to disrupt Britain's top companies' annual general meetings. By buying shares in the companies, which included BP, Shell, BAE, and Rio Tinto, activists were able to publicly hold board management to account at the AGMs on issues ranging from bribery to environmental destruction to mistreatment of third world workers.

At a time when the threat of hostile takeover is ever present, and competitive pressures are ever greater, corporations must handle their shareholders' concerns and demands carefully. Politics is buffered by its monopoly status, but business is increasingly vulnerable to the vagaries of the market.

Beautifying the Brand

In the age of the logo, reputation is paramount. Corporations are increasingly realizing that there are new expectations of them. As their actions become more and more public, they are expected to justify their policies and actions, and address their consumers' and shareholders' concerns, to a hitherto unprecedented degree. Does the company mistreat its employees? Is the company damaging the environment? Is the company backing a repressive regime? Can the company be trusted? These are increasingly asked questions, from key stakeholders rather than by politicians.

No wonder, when such priorities are stated, that the World Economic Forum chose as the theme of its 2003 conference "Rebuilding Trust," and that many corporations are professing great concern. CEOs of several leading brand-name companies confess to the vulnerability that they feel under the media spotlight. "What we fear most," says one, "is not new legislation, but consumer revolt." Another speaks of a feeling of powerlessness: "If people think corporations are powerful, they haven't been in a corporation. We are by no means powerful—we are confined and restricted in what we do. Consumer choice doesn't allow us to have unfettered power."[20]

Although still a relatively new phenomenon, the forcefulness of today's consumer and shareholder activist movement, and the breadth of its "church"—especially in the wake of the Enron, Adelphia, and Tyco debacles—have triggered a response from corporations or, at least, from those who are caught. While governments have allowed the interests of corporations increasingly to take precedence over the public interest, the public, through the marketplace, is reclaiming its interest for itself.

It is too early in this process to gauge the size of the impact. Only a few corporations have been actively targeted; corporations that do not sell consumer goods are to a large extent shielded from activist action, for highly visible brand names obviously provide a better target for smear campaigns and other public attacks;[21] and instances of active protest are still relatively few, so that each gains significant media interest.

However, in the same way that governments are now unwilling to risk raising taxes because of the backlash from voters, many captains of industry have a fear of adverse publicity.[22] The cost of rebuilding a negative reputation is high; in fact, once lost, a reputation may be impossible to regain. Memories of the GM debacle, Brent Spar, and Kathie Lee Gifford have left their mark.

Increasingly, the cost advantages of cheaper labor or cheaper inputs from suppliers dismissive of human rights, or actuarial calculations of the risks of retroactive environmental rulings, must be weighed against the damage from negative publicity, the cost of poor public relations, and the possibility of consumer protest. "The spotlight does not change the morality of multinational managers; it changes the bottom-line interests of multinational corporations."[23] Bottom-line interests that politicians seem increasingly unable to meet.

Many corporations are now concerned with ensuring that they are well thought of by the public and the media. In practical terms this has meant a substantial increase not only in the number of multinationals that now have codes of conduct (*all* Fortune 500 companies in the United States do), but also in those now willing to undergo an audit of their environmental, and to a lesser extent social, policies. Whereas in 1993 only 15 percent of Europe's largest companies had carried out environmental audits, and social audits were the domain of the ethical brands like Ben & Jerry's and the Body Shop, by 2000 more and more companies were taking this tack, with over half carrying out environmental audits and a number of significant players such as Shell, BP, GrandMet, and BT now carrying out social audits, too.[24]

While the independence of the auditors remains in some cases questionable, and companies' willingness or indeed ability to respond to findings is still unproven, what is clear is that corporations are acting more transparently than before. The rhetoric of business, at least, is changing as many of the biggest brand names seem to have realized that while they do not always have to be seen to do right, a preemptive admission of culpability is probably a better PR strategy than a denial or even a valid explanation, once they have been exposed in unethical practices.

Unlike parties or politicians that have been "outed," most of the companies and industries that have been singled out by activists have made changes in their way of doing business. Nike initially adopted a very defensive posture in late 1997 when environmental researcher Dara O'Rourke revealed that poor ventilation, exposure to hazardous chemicals, and inadequate safety equipment and training at a plant in Indonesia were putting the health of workers at risk. But a year later Nike allowed O'Rourke back into the plant to establish, and presumably publicly report, that changes promised by the company had actually been made. Global Exchange, a human rights organization dedicated to promoting social justice around the world described this as "an astounding transformation for a company that once treated independent monitoring as a public relations exercise." Furthermore, the company claims to have severed contracts with plants that paid below minimum wage levels, promised free after-hours education for all its third world factory workers, and raised the minimum age of footwear factory workers to eighteen and the minimum age for all other light manufacturing workers to sixteen. If Jeff Ballinger of Press for Change, a group which monitors workers' rights in Asia, was right when he said that market research revealed that twelve-year-old girls in focus groups back in the States were talking about Nike's labor abuses, it is not surprising that the company acted in this way.

Thanks to the boycotts of sporting goods in the mid-1990s and resultant negative publicity, many footballs in America now carry a label declaring "No child or slave labor used on this ball." Only a few years ago, Disney wouldn't have considered sweatshop workers in Indonesia to be important stakeholders in their business; now they have no choice but to do so. The diamond industry, chastened by the experiences of the antifur lobby during most of the 1990s and alerted to the threat of a consumer boycott, finally woke up to the fact that consumers might not want to wear gems that were funding the apparatus of war; diamonds are a girl's best friend no longer unless they are seen to be "blood free."

Nearly all the companies targeted by the joint action of the Free Burma Coalition and associated consumers and shareholders pulled out of Burma. Before governments and state authorities proved willing to take action against the repressive regime, consumers themselves applied sanctions with some success. Siding with Aung San Suu Kyi rather than politicians or corporations on the question of the effectiveness of such action, Generation X turned its back on the "choice of a New Generation," joined forces with the baby boomers, and marched with the Free Burma Coalition in rallies that targeted well-known brands. Philips, Heineken, C&A, Ralph Lauren, Motorola, Carlsberg, and Kodak were just some of the multinationals who decided that the potential attractions of the Burmese market were not nearly great enough, and they left the country, fearing a damaging boycott were they to remain.

These are just a few examples of real changes that corporations have made over recent years in direct response to actual or threatened consumer or shareholder revolt. The corporations did far more than merely comply with their legal obligations. Neither legislation nor regulation drove these firms to change the way they did business—in most of these cases government or regulatory demands have been weak or absent—but consumers who, by boycotting products or exerting pressure at shareholders' meetings, are beginning to make companies realize that responsible behavior is increasingly less of an option; rather it is a necessity if a company means to continue to safeguard its commercial interests. Many of these consumers are largely uninterested in politics, believing that politicians do not listen to them and court only big corporations. Until politicians learn to respect and trust the electorate, voters will not be inclined to reciprocate.

The Impotence of Politicians

So while consumers are beginning to play the role of global policemen, what are politicians doing? Afraid to resist corporations directly, they are tacitly endorsing consumer activism. Politicians are stepping aside so that consumers and shareholders can become, to an ever greater extent, the watchmen over corporations' activities.

It is increasingly evident that concerns are being inadequately represented in the traditional political sphere and cannot be met through public sector provision. The growth of environmentalism in the 1980s, for example, was a response to the failure of political parties to take seriously the environmental costs of industrial production.

Environmentalism has now moved into the mainstream, but despite the "greening" of politics, few governments have succeeded in challenging the vested interests of the business community and imposing a reduction of carbon fuel emissions or the safeguarding of natural resources. Environmentalists' response has been to target multinational corporations and international organizations directly. In a dual strategy they have fought a PR battle against big business, exposing its record of destruction, while battling at local levels to prevent environmentally damaging developments. In the midst of confrontations and protest, national governments have been marginalized, alienating both sides by endorsing proenvironment rhetoric while using the coercive power of the state against protesters—or, as in the case of Monsanto, actively coming out on the side of the corporation.

Politicians, unable and unwilling to take corporations to task for fear of jeopardizing their relationship with them, and recognizing their own inability to effect change on a global stage that lies outside their jurisdiction, are furtively supporting consumer vigilantism and imitating its language and demands. In the United States former president Clinton urged companies to be "good corporate citizens" by monitoring working conditions at manufacturers which produce their goods.[25] In the U.K., Trade and Industry Secretary Stephen Byers endorsed the new culture of consumer power, calling on individuals to use their wallets and voices to protest against high prices and end the image of "Rip-off Britain."[26] The British Department for International Development is encouraging retailers to adopt buying policies that would satisfy nongovernmental organizations (NGOs) such as Greenpeace and Amnesty International.

But these requests and urgings are rarely backed up by the muscle of the law. The British government asked Premier Oil to pull out of Burma but did not *make* it do so (it did not) by imposing economic sanctions on Burma. Companies are not compelled to enforce labor standards in their overseas operations, under threat of penalties at home, despite politicians' rhetoric. External verification of environmental programs and reports is not required either in the United States or in most of Europe. Governments issue ethical declarations but fail to back them up with legislation. For all Clinton's insistence that America's trading partners improve their record on human rights, and the British government's talk of an ethical foreign policy, these proved to be empty promises in the face of possible loss of trade. The only financial penalties faced by companies that trade with countries such as Indonesia, China, and Burma are the consumer boycotts imposed by Western shoppers.

Many politicians recognize their own impotence. Edward McMillan Scott, leader of the Conservatives in the European Parliament, tells a story that clearly illustrates the helplessness that many politicians feel. In November 1996 he organized a rally with Ken

Wiwa, son of Ken Saro-Wiwa the Nigerian activist and critic of Shell's operations in Ogoniland, to commemorate his father's execution. Crowds gathered and someone asked him, "Mr. McMillan Scott, what are you going to do about this injustice?"

McMillan Scott replied, "I'm going to stop putting Shell petrol in my car." Taking action as a consumer was, in his opinion, more effective than anything that he could do as a politician.

7

All That Glitters . . .

Outfoxing the Fox

At first it all looked so promising. Husband-and-wife team Steve Wilson and Jane Akre joined Rupert Murdoch's Fox 13 TV station in Florida trailing clouds of glory. She was a former CNN anchorwoman, he had won four Emmy awards for investigative journalism. His documentaries on Chrysler's faulty door locks, corrupt senators, and Ford's fire hazard ignition had made him one of the most famous and feared journalists in America. They had been hired to beef up the Tampa Bay station's serious news coverage.

Yet within a year both had been fired. Their only crime, they claimed, was their refusal to broadcast untruths about agrochemical company Monsanto, a major Fox advertiser.

The couple decided to fight Fox in the courts under Florida's whistleblower legislation, alleging that they had been sacked for simply trying to tell the truth. And, in August 2000, while the jury did not find for Wilson, Jane Akre won.

The David and Goliath nature of the court case afforded a disturbing insight into the complicated relationship between major advertisers and media moguls in the United States. "We were telling the secrets the supermarkets didn't want aired, the dairy farmers didn't want, and Monsanto didn't want," says Wilson of the trial. "When you put those advertisers together, you're talking about a lot of advertising dollars, both in Tampa and at Fox's other stations across the country. . . . When media managers who are not journalists have so little regard for the public trust that they actually order reporters to broadcast false information and slant the truth to curry the favor or avoid the wrath of special interests, as happened here, that is the day any responsible reporter has to stand up and say, 'No way!' "

The story begins in 1996, when Rupert Murdoch bought thirteen major TV stations in America to add to his Fox cable network. This brought his holdings of American TV stations to twenty-two, reaching more than half the nation's viewers. Tampa Bay's WTVT, a former CBS station known for in-depth reporting, was one of the new acquisitions. High-profile duo Akre and Wilson were brought in to boost ratings; initially it looked as though the station would maintain its high journalistic standards.

The pair were quick to uncover a major story. They discovered that Florida's milk supply came from cows injected with a substance called bovine growth hormone, or BGH. Sold under the brand name Posilac, it can boost a cow's milk yield by as much as a third. It is made by U.S. agrochemical giant Monsanto and was approved by the FDA in 1993. However, BGH was, at the time, banned in Canada, Britain, New Zealand, and most of Europe. Scientific research has suggested that the hormone could be linked to cancer, a theory strongly disputed by Monsanto. In response to consumer fears in Florida about BGH, the state's giant supermarket wholesaler of milk and dairy products announced in February 1994 that it would not buy milk from treated cows "until there was widespread acceptance" of the hormone. But in 1996 Akre photographed cows receiving Posilac injections at all of seven randomly selected Florida dairy farms, challenging the milk supplier's claims.

WTVT, by then known as Fox 13, was enthusiastic about the story. The station booked several thousand dollars' worth of radio advertising to promote the documentary, scheduled for broadcast on February 24, 1997. But days before transmission Monsanto's lawyers contacted Fox's head office in New York, claiming that the documentary was inaccurate. The Monsanto letter of complaint included a sentence which Akre and Wilson found particularly disturbing. It read: "There is a lot at stake in what is going on in Florida, not only for Monsanto, but also for Fox News and its owner." The documentary was pulled "for further review."

Initially Fox 13 refused to back down: the news manager examined the film and, according to Akre and Wilson, found no reason to doubt their claims. A new date was set for a broadcast the following week. But then Monsanto's lawyers again wrote to the Fox head office, saying that the reporters were biased and that the story would damage the country. The station pulled the BGH broadcast again. It was never to be shown.

Shortly afterward, Fox 13's news manager was fired. According to the journalists' affidavit, the new management offered Akre and Wilson "large cash settlements" in exchange for their resignations and a promise not to publish details about Posilac or how Fox had handled the story. They refused. Over the following six months they rewrote the story seventy-three times, at Fox's insistence, but consistently refused to include an assertion that they strongly believed to be false: that BGH milk is as safe as milk from untreated cows. Finally, in December 1997, the pair were dismissed for "insubordination."

Akre and Wilson decided to sue Fox for violating Florida's "whistleblower" act. "We are parents ourselves," Akre said at the time. "It is not right for the station to withhold this important health information. Solely as a matter of conscience we will not aid and abet their effort to cover this up any longer. Every parent and every consumer has the right to know what they're pouring on their children's morning cereal."

In court Fox insisted the dismissal had "nothing to do" with the BGH story or with letters from Monsanto. David Boylan, Fox 13's general manager, said the pair were fired for "contentious, argumentative, *ad hominem,* and vituperative conduct and for refusal to abide by [Fox 13's] established policies and procedures."

But on August 18, 2000, after a five-week trial and six hours of deliberation, a Florida state court jury found that Jane Akre's threat to blow the whistle on Fox's misconduct to the Federal Communications Commission (FCC) was the reason for her termination, and it awarded her damages of $425,000.

"We set out to tell Florida consumers the truth about a giant chemical company, and a powerful dairy lobby clearly doesn't want them to know," Wilson said. "That used to be something investigative reporters won awards for. As we've learned the hard way, it's something you can be fired for these days whenever a news organization places more value on its bottom line than on delivering the news to its viewers honestly."

The Truth Will Not Be Televised

In the previous chapter I may have risked giving the impression that consumer politics operates in some approximation of a perfect information environment. The Fox story shows that this is not the case. Consumers and shareholders are often left to operate blind. Their inability to get full and accurate information is a major handicap for activists.

In the world of print journalism boundaries are blurred as the "walls" between advertising and editorial departments come tumbling down. The *Los Angeles Times,* for example, was exposed as having devoted the October 1999 issue of its magazine to coverage of a new Staples Center sports arena, under a deal to share revenues with the center. "A flagrant violation of the journalistic principle of editorial independence," wrote David Shaw in the newspaper's own postmortem.

News broadcasting is also becoming increasingly commercialized as the pressure on broadcasters to make profits continues to increase, creating an overriding focus on ratings and advertising revenues. Once again, the pressures are greatest in the United States, but other countries are following fast. Belgium is the only European country where television still remains free of commercials. In the U.K., as David Liddiment, director of programs for ITV Network has put it, "We are less able to ignore the commercial imperative than ever before, although we're not running an audience delivery service for advertisers. Our job is to provide a service for viewers that serves the advertisers at the same time."[1]

Of course, serving the best interests of the advertisers may not be in the best interests of viewers. Fox's story dilemma is faced by an increasing number of broadcasters: whether to run a story, even if it risks jeopardizing crucial advertising dollars.

Recognizing the explicit corporate agenda we tend to be skeptical of the claims of product labels and advertising copy but are less likely to scrutinize the traditional sources of our news, which we assume are free from external influence. But for every Brent Spar, child labor, or GM disclosure that gets on the air, how many others, like the BGH story, are being suppressed?

Rupert Murdoch is, of course, notorious for using his control of information to support his business interests. As is by now well known, in his quest to capture a substantial slice of the Chinese media pie, Murdoch worked hard to gain the favor of the Chinese authorities. In 1994 he dropped the BBC World Service from his Asian Star TV satellite after it criticized Chinese leaders for the Tiananmen Square killings—the BBC's independence and impartiality were potentially damaging Murdoch's prospects in the Chinese market. And in 1998 HarperCollins, a Murdoch-owned publisher, cancelled publication of *East and West,* a book by Hong Kong's last British governor, Chris Patten. It was clear that publication risked aggravating the Chinese authorities given that Patten's recollections of his time as governor of Hong Kong were highly critical of the Chinese government.

The suppression of stories that may harm corporate interests is hardly unique to Murdoch's media empire. A segment that was aired on NBC's *Today* show about defective bolts in nuclear plants reportedly omitted the following lines: "Recently General Electric engineers discovered that they had a big problem. One out of every three bolts from one of the major suppliers was bad. Even more alarming, GE accepted the bad bolts without any certification of compliance for eight years."[2] General Electric owns the network. Similarly, Disney-owned ABC quietly dropped a report which alleged that pedophiles had been employed at a Disney theme park.[3]

When tobacco advertising was still allowed on American television, a clear correlation was found between the amounts of money the networks derived from tobacco advertising revenues and their willingness to enter the debate on the health effects of smoking.[4] In Italy where two influential newspapers, *Il Messaggero* and *Il Tempo,* are owned by construction companies, "there is plenty of auto-censorship by journalists keen not to upset their bosses."[5] Liza Brinkworth, the British investigative journalist who exposed tales of underage sex and drug abuse at the Elite model agency, failed to sell her story to numerous women's magazines—they were clearly not prepared to risk fashion-related advertising. It was the publicly funded BBC that eventually broke the story.

Either to avoid jeopardizing relationships with advertisers or to safeguard their wider interests—increasingly these media organizations form parts of larger groups of companies, with business interests outside the media—news reporting has frequently been found to be skewed in recent years. This is not, of course, a new phenomenon. "A study of women's magazines in the period 1983 to 1987 revealed that not one magazine that carried cigarette advertising published any full-length feature, column, review, or editorial on any aspect of the dangers of smoking. During the same period lung cancer was determined to be the number one killer of women, surpassing even breast cancer. Not one of the magazines surveyed mentioned this fact."[6]

What is new, however, is that there has been a consolidation of the media industry over the past few years, and as few as ten global media players now wield immense power. As a result, the regulation of broadcasting has been much weakened. This is particularly the case in the United States where, for example, in August 1999 the FCC did away with longstanding media ownership rules that forbade the largest television companies and networks to own more than one station in the nation's largest cities; and where a federal plan to license hundreds of new noncommercial low-power stations throughout the country was, in December 2000, effectively killed off by a huge lobbying campaign launched by big media interests. What we do or do not learn, at least through traditional channels, will increasingly depend on the decisions of a very few.

Aidan Whilee, general secretary of the International Federation of Journalists, the world's largest organization of journalists representing more than forty-five thousand members in 103 countries, said when the $350 billion Time Warner–AOL merger was announced: "We are now seeing the dominance of a handful of companies controlling information and how that information reaches people. Unless action is taken to ensure journalistic independence, we face a dangerous threat to media diversity. . . . Otherwise we will have corporate gatekeepers to the flow of information, who will define content to suit their market strategies."[7]

As consumers of news, we are unable to police the news providers in the way we police other corporations. Unless the media are held accountable to an external and independent force, our independent press, one of the vital components of democracy, may be in danger.

Seeing the Unseen

In a world in which governments are increasingly unable to keep corporations in check, and consumers and shareholders have to take on this role, loss of a free press is devastating. For if a story is not told, if a problem is not seen to exist, people have no reason to protest.

But while it is likely that such scandals still usually—and eventually—come to light, given the media's attraction to David and Goliath stories (which also greatly appeal to the public), even when they do come out, consumers, faced with the subsequent public relations onslaught, find it increasingly hard to gauge which sources of information can be relied on.

Consumers in the U.K. who sought a realistic assessment of the risks of genetically modified food, for example, were left to piece together the evidence on the basis of hard sell from the agrochemical companies, halfhearted reassurances from governments, hysterical press reporting, and passionate denunciations by environmental groups. The findings of scientists who claimed to have rigorously tested the effect of GM foods were disputed by other scientists. In this, as in many other cases, it became impossible for consumers to act on the basis of objective evidence. They simply did not know whom to believe.

In the United States, on the other hand, most people didn't even know that their foods were being genetically manipulated. While a 1999 *Time* magazine poll found that 58 percent of Americans would avoid genetically engineered foods if they were labeled, half of Americans surveyed in 2000 thought that their food was free of biotechnological manipulation. In reality more than 60 percent of processed foods sold in America are GM or contain GM ingredients.

Even for those who wish to actively seek out the truth, dealing with the range of potentially conflicting information, not to mention misinformation, is both time-consuming and confusing. Whom to believe? Whom to trust? Which issues to champion? Is it Nike, Reebok, or Adidas that I shouldn't be buying? With so many contradictory stimuli, and so many brands, even the most concerned consumers face information overload and compassion fatigue. Simply increasing the amount of information does not solve the problem, especially if the sources are tainted. Without a reliable source, the big consumer battles are likely to be won by those who shout the loudest.

When stories break, the media's interest is fleeting. Consumer campaigns are highly dependent on media coverage, but the media are by nature short-termist in their outlook. Their intense but brief attention to most political protests rarely reflects the longer-term nature of the issues. Like pressure groups, they have a vested interest in creating a hysteria that stimulates interest and sells more newspapers. During 1999 the campaign against GM foods went from being the preserve of left-wing environmentalists to a tabloid crusade and has now fallen back to relative obscurity.

Once something is no longer reported, people inevitably start to think the issue has gone away. For all except the most committed, boycotting of products tends to be short-lived, and once the issue is less newsworthy consumers generally revert to their original preferences, unless their chosen substitute has actually turned out to be as good or better.[8]

Although consumer activism is undoubtedly entering the mainstream, campaigns which are high-profile, intense, and fashionable are understandably more attractive and successful than those that are more routine and prosaic, however worthy of support. In recent years it has been possible to witness the fashion trends of consumer protest: from Nestlé boycotts in the late 1970s, through antiapartheid protests in the early 1980s, global warming and rainforest depletion later in that decade, live animal exports in the early 1990s, then the rights of workers in developing countries, and most recently food safety. At each stage protesters secured small victories and corporations changed tack, but with the exception of apartheid, in no case was the war won so conclusively that further protest was unnecessary, as we saw with the repeated denunciations of sweatshop labor. Consumer activism seems most effective when consumers remain active.

Rhetoric or Reality

The haphazard information provided by the media is contrasted by the stream of information produced by companies. Corporations are spending more than ever before in marketing positive images of themselves. Of course, company-provided information is usually treated with at least a degree of skepticism; it is only to be expected that a company will trumpet its successes and try to cover up its wrongs, and it will typically tell only one side of the story, the side it wants its customers to hear.

The British supermarket chain Sainsbury's, for example, runs advertising campaigns claiming that it is pro-organic and anti-GM, in order to appeal to the prevailing public mood, while at the same time it is quietly involved in the development of new genetically modified strains of vegetables that will have a longer shelf life and reduce wastage. Dr. Philip Dix, from the National University of Ireland in Maynooth near Dublin where the research is being carried out, has said, "[Sainsbury's] is taking a passive role because of the climate [toward GM foods]. It prefers not to be too closely associated with the project."[9]

As we saw in the case of Monsanto in the previous chapter, marketing spending is not always the solution. The firm misguidedly attempted to counter European consumer fears in 1999 with a $1.6 million advertising campaign proclaiming the merits of genetically modified foods. The slogan on its website just before the crisis was "Food, Health and Hope." Hope it's right? Hope it works? Hope the public buys it? In this case the public didn't. The campaign backfired, and European consumers and green lobbies created enough pressure to ensure that grocery retailers and restaurants were cautious about stocking and using GM products. The public is increasingly unlikely to take such platitudes at face value when they are posted on company websites or found in the companies' promotional materials.

What the public is less immediately able to write off as mere marketing, however, is the vogue for codes of conduct and environmental audits to which, as we saw, most leading companies now subscribe. All Fortune 500 U.S. companies and over half the U.K.'s top five hundred companies, for example, now have codes of conduct.[10] In some cases they clearly have real significance: according to Sir Martin Sorrell, CEO of WPP, one of the world's largest communications groups, managers throughout his thirty-three-thousand-employee organization who are found not to be adhering to the code are fired. However, a survey from the Institute of Business Ethics reveals that such codes are often not "active documents," but lie dormant in company filing cabinets.[11]

Although over half of FTSE and Fortune 500 companies now carry out environmental audits, a study of environmental reporting by the oil industry concluded that it remains "virtually impossible" for stakeholders to draw meaningful comparisons between firms.[12] Only 18 percent of the environmental reports of the top one hundred firms in Australia, Belgium, Denmark, Finland, France, Germany, the Netherlands, Norway, Sweden, the U.K., and the United States surveyed by KPMG were found to have been independently verified. Even then the study concluded that "verification is a long way from a standard which readers can rely on to guarantee the reliability of the reported data and information," and "users still have to read between the lines in order to interpret the report and the verification statement."[13]

With no external means of verification or political or legal enforcement (though some Scandinavian countries, the Netherlands, and Australia have introduced limited reporting requirements under "green accounts legislation"), the accuracy of company reporting will continue to be difficult to ensure, a problem which will stand firmly in the way of consumer empowerment. In October 2000 the BBC current affairs program *Panorama* revealed that Nike and Gap were still using suppliers who were employing underage labor despite their claims to the contrary. Even the trailblazing ethical retailer the Body Shop was accused of inaccuracy over its initial pledge that its products were not tested on animals, and it subsequently substituted the watered-down slogan "against animal testing."[14] Sheila McKechnie, director of Britain's Consumers Association, was among those pressing for a Freedom of Information Act that would give customers better access to the information they need to wield their power as consumers more effectively and appropriately. Unfortunately the version of the act that was ultimately passed was so weak that it barely affected the information asymmetry between consumer and corporation.

It is even more confusing when claims are made or endorsements provided by supposedly neutral third parties, who then turn out to have hidden corporate interests. An extreme case of this was Nestlé, which in the late 1970s employed marketing representatives to dress up like doctors in order to sell the company's baby milk to mothers in Africa as the healthiest alternative for their children. While deceptions such as these remain thankfully rare, what is increasingly commonplace is a blurring of the boundaries between independent research and the corporate agenda.

In the United States, corporate funding of academic scientific labs more than doubled over the ten years to 1997, when it stood at $2 billion.[15] In Britain scientific public research money fell by 20 percent between 1983 and 1999, leaving a deficit that corporate funds were more than prepared not only to fill but also to augment. Although these boosts to research budgets can clearly be beneficial, conflicts of interest can, unsurprisingly, arise. Samuel Cohen, a University of Nebraska researcher on saccharin, whose findings were heavily relied on by the U.S. government in justifying its decision to take saccharin off the list of cancer-causing chemicals, was revealed to have been funded in part by an industry group whose members included Cumberland Packing, the makers of Sweet 'n Low saccharin products. Exxon Mobil has provided funding for maverick scientists who claim there is insufficient evidence of a human factor in climate change. In 1998 the company donated $10,000 to the science and environmental policy project run by Fred Singer, a highly vocal critic of the global warming theory, and also gave $65,000 to the Atlas economic research foundation, which promotes Singer's work as offering "a wealth of information, credibility, and encouragement." Particularly worrying given that George W seemed to use these views to justify his rejection of Kyoto, claiming that the scientific work of global warming was still "unsettled."[16] And Bush's regulation czar, John Graham, solicited $25,000 in funding from Philip Morris at the same time as he was overseeing a study that concluded that there were no health risks from secondhand cigarette smoke.[17] "I feel academia is becoming tainted in this," says Drew Pardoll, an oncologist at Johns Hopkins University School of Medicine. "It's an issue of public trust."[18] If scientists will not publish in medical journals for fear of losing out financially, or if findings are suspect because of potentially divided loyalties, our ability to access impartial scientific research in many areas will be lost.

It is not just the impartiality of scientific research that is being questioned. University centers, think tanks,[19] public interest organizations, consumer organizations, and even religious leaders are now funded by corporations. Nottingham University announced in December 2000 that it had accepted £3.8 million from British American Tobacco, which at the time was under investigation over smuggling allegations, to finance a new school of, most ironically, "corporate social responsibility."[20] In the United States, the National Consumers' League, which describes itself as "America's pioneer consumer advocacy organization," got 39 percent of its income in 1997 from corporations and industry associations. "Almost every current project, seminar, brochure, newsletter, and fund-raising dinner is sponsored in large part by major corporations or industry associations."[21] These donations clearly risk jeopardizing impartiality. And according to the World Health Organization, Philip Morris sought to identify and encourage support for Islamic religious leaders who opposed interpretations of the Koran which would ban the use of tobacco.[22] Is nothing sacred in the world of the Silent Takeover?

Even the allegiance of regulatory agencies has been questioned. Three of the key figures in the Food and Drug Administration who were responsible for approving the BGH hormone had ties to Monsanto. One was a former Monsanto research scientist who had worked on BGH while at the company; another had been a lawyer with King & Spalding, a firm that represented Monsanto, and had helped draft regulations to be used by local governments to fight labeling of milk from BGH-treated cows; and the third had worked on Monsanto-funded studies at Cornell University.[23]

Although a subsequent investigation found no impropriety, Bernie Sanders, the Independent congressman who requested the investigation, remained adamant that corporate interests had been at play. "The FDA allowed corporate influence to run rampant in its approval of the drug. The ethics rules were often stretched to the breaking point and broken on a number of occasions," he said.

E-Activism

In this confusion of information and misinformation from the traditional media, governments, corporations, think tanks, and research institutes, new sources of information have become essential to activists. We saw in the previous chapter the effort NGOs and pressure groups are now putting into getting their point of view across—with real success. In a 1996 survey of public confidence in various sources of information about modern biotechnology, respondents trusted consumer and environmental organizations most, at 30.5 percent and 22.4 percent respectively. Only 7.8 percent put their trust in public authorities and 1.6 percent in the industry itself.

But what has really revolutionized information over the past few years, and has opened up completely new sources of information for all of us, is of course the Internet. Most of this information, at least at present, is out of the control of corporations or large organizations. Information provision is no longer the domain of the media giant. Any individual or organization, with a minimum of technical know-how and equipment, can tell the world whatever they want it to hear by creating their own website.

Those who are looking to monitor corporate activity have never had an easier time of it. Type "McDonald's" into any search engine and you will quickly be directed to the "I hate McDonald's" site—the voice of a single disgruntled customer—and the McSpotlight site at www.mcspotlight.org, which gets over one million hits a month. This site not only provides full background on the infamous "McLibel" trial, at the end of which in March 1999, three British lord justices ruled that it was "fair comment" to say McDonald's workers worldwide suffer poor pay and conditions, but also accuses the company of doing all kinds of harm, from environmental destruction to the "McExploitation" of kids.

Go to the Corporate Watch site at www.corpwatch.org, and you can "find resources designed to help you find out more than you probably wanted to know about transnational corporations" and obtain advice on how to "dig up the dirt on your favorite corporation." On its Nike page, for example, you can read Ernst and Young's confidential November 1997 labor and environmental audit of Nike's facility in Vietnam, examine the lawsuit brought against Nike for presenting an allegedly false picture of their working conditions, peruse a collection of news articles on Nike's operations, and look at photos taken inside a Nike plant in Vietnam—including pictures of workers using dangerous materials, such as glue and solvents, without appropriate protective gear.

Don't like Bill Gates? There are numerous sites that have taken up the battle against Microsoft. NetAction, a U.S.-based nonprofit organization, features reports and activist resources for "fighting the Microsoft Monopoly" and provides a digest of anti-Microsoft websites. At www.usdoj.gov you can find the United States Department of Justice's legal documents for its antitrust case against Microsoft. Elsewhere you can find a memo leaked by a Microsoft whistleblower, describing an internal management plan to prevent consumer groups and state attorneys general from pursuing antitrust action against the company.

The range and quality of such sites is immense. And as quickly as companies try to close them down, they reappear in other guises. A recent development has been that of corporations buying the domain names of oppositional sites. Domino's pizza bought ihatedominopizza.com; and Chase Manhattan Bank has acquired the rights to ihatechase, chasestinks, and chasesucks. But this strategy seldom works; there are just too many permutations. Scott Harrison, a twenty-three-year-old New Yorker, launched a site against Chase in protest of an erroneous bill of $650 that it took him seven months and thirty phone calls to correct and called it chasebanksucks.com. Hardly activism at its most noble.

But the Internet does not only provide information passively.

What most differentiates the Internet from traditional media is that it is an interactive medium. Community boards and newsgroups allow members to share stories and complaints. Wal-Mart Watch (www.walmart.watch.com) actively solicits personal stories relating to Wal-Mart's impact on local business, employees, and consumers which are then posted on its site. Wal-Martyrs (www.walmartyrs.com) asks Wal-Mart employees, or former employees, to share their experiences and post their stories. And chatrooms provide live forums for discussing companies.

The Internet is like a game of telephone multiplied and magnified. It provides the ultimate medium for conspiracy theories; rumors pass across borders and time zones almost instantaneously. Messages appear onscreen promiscuously, and there is no easy way to separate the truth from lies. From companies' point of view, this aspect of the Internet is chipping away at the significant benefits they are deriving from the dot-com revolution. It is proving to be a corporate nightmare, a medium which, to quote the CEO of one of the world's largest car manufacturers, "promotes half-truths and irresponsible representations of their companies that cannot be controlled or influenced"[24]—although they are, of course, trying to do so where possible. Many corporations are now hiring third parties such as eWatch to monitor anticorporate sites for libel and to help with damage control when criticism of their companies is posted on the Net.

From the consumer's point of view, however, the Internet provides a means to scrutinize corporations more directly than ever before and unprecedentedly easy ways of taking action against them. There are sites that provide the social and environmental ratings of companies,[25] sites that provide information on boycotts,[26] sites that give standard form letters of protest to CEOs, sites encouraging people to write to corporate moguls, and sites telling them which stores to picket.[27] The Essential Action site (www.essential.org) was, in its own words, "created to alert activists to current international campaigns and activities." Current campaigns include an onslaught on the tobacco industry; race discrimination lawsuits against Coca-Cola; and "Boycott Shell/Free Nigeria." The Boycott Board's stated purpose is "to provide the socially conscious consumer with a means of learning about various boycotts in progress."

Search under "child labor" on www.directhit.com, and the seventh site that you are urged to visit recommends that you:

> Ask the right questions
>
> By making store managers and corporate officers aware that you have concerns, you encourage them to take action. Does their company guarantee products being sold are made under humane conditions? Can you get a list of names and addresses of contractors and subcontractors? Does the company have a code of conduct? Can you have a copy?
>
> Visit stores
>
> Retailers have a saying that "the customer is always right." They may not actually believe that, but it does point out the importance of customers' opinions. After all, without customers, there's no store! Ask your store manager questions, and make sure they know you want honest answers. Often, organized visits by groups of customers can produce immediate results as chain store managers will call head office to get the answers.

This site is not that of a radical fringe organization, but of the UCLWA, one of America's largest trade unions.

It is uncertain how long the Internet will be able to retain its irreverent and antiestablishment leanings. Takeovers, corporate influence, and control, and commercialization are all likely in time to threaten its democratic and egalitarian spirit. Microsoft continues to steer Net users to its own websites and those of its commercial partners.[28] Time Warner can now direct a torrent of information to AOL's 22 million subscribers.

But that does not mean that the traditional media will ever be able to rule cyberspace. Despite the head start that traditional media companies have in terms of available funds and avenues to promote their websites, the explosion in socially responsible sites, portals, and public forums over such a short period means that in all likelihood many of these alternative news sites will not only remain intact, but will actually become larger and more powerful. Disinfo.com receives eleven thousand visits a day, indymedia.net gets more than one million hits a month, and www.Greenpeace.org was receiving over fifty-eight thousand visitors a week by mid-2000, up fourfold from four years earlier.[29] Type in "Nike boycott" on a typical search engine and you get a list of over six thousand sites.[30] E-activism on the Net continues to rise.

The number of people who actively seek out this kind of information may be relatively small, but once their findings are passed on to others the numbers grow exponentially. E-mail allows people to send news to hundreds of others almost effortlessly. Each of these, in the same way, can pass the story to hundreds of people on their address lists—the e-mail correspondence between Jonah Peretti and Nike over his request to have Nike stitch the word "sweatshop" onto his customized trainers was forwarded to over a million people. Add this to the increasing ease of communication all over the world with cell phones, satellite pagers, and so forth, and the viruslike infectivity of today's news is clear, not to mention the impossibility of containing it. Where governments have started to crack down and regulate the Internet, as has happened in Vietnam and Burma, ways have been found to bypass attempts at controls; for instance, accessing the Internet illegally by dialing out of the country using a cell phone. "By jumping over borders, by opening cheap access to information, and by providing forums for debate in countries where the media are monopolized, the Internet offers the disenfranchised a chance to participate."[31]

The pervasive urban myths of the 1980s and 1990s—the rat found in the Kentucky Fried Chicken carton, Marlboro's relationship with the Ku Klux Klan (the secret sign was revealed by folding a Marlboro cigarette pack in a certain way)—miraculously managed to circle the globe even in the pre-Internet era. Now such things are commonplace. Truths, half-truths, and blatant lies cross national borders with ever-increasing speed.

. . . Is Not Gold

Of course, we should be extremely cautious in assuming that alternative sources of information are necessarily any more socially responsible than the traditional media. Pressure groups, aware of the importance of coopting consumers in their battle against the corporations, have been found on several occasions to exaggerate risks. In 1995 the Advertising Standards Authority, the U.K.'s advertising watchdog, accused pressure groups—including Friends of the Earth, Greenpeace, and the International Fund for Animal Welfare—of exaggerating claims, exploiting public trust, and damaging "the credibility of the advertising industry as a whole."[32] All of these groups had used shocking or misleading advertisements. Greenpeace overestimated by a factor of thirty-seven the amount of hydrocarbons the Brent Spar oil platform might leak into the sea. And "when the Braer oil tanker went aground off Shetland in 1993 and spilled tens of thousands of tons of crude oil into the sea, wildlife groups predicted catastrophic effects on marine life which were never borne out."[33] Pressure groups' need to influence public debate often provokes them into the creation of unwarranted public anxiety.[34]

The drive to win media attention may also reduce the ability of groups to focus their campaigns effectively. Many environmental groups have been found to be "keener on getting the attention of the media than on devising a sophisticated political strategy."[35] Raising popular awareness and support must be a fundamental part of all consumer campaigns, but it remains true that successful protests usually also require political strategies directed at policy makers or senior politicians. Neglect of these aspects of a campaign decreases its effectiveness.

And, of course, consumer pressure groups have their own agendas and priorities. What determines the malpractices on which they choose to focus? What motivates them to protest? Can we assume that they are any more high-minded than corporate executives? Whose interests are they really safeguarding? For just as corporations usually only promote ethical causes when it is in their interests to do so, the same is often true of consumer groups. This self-interest may be narrowly conceived as the interests of the group members—for example, an unwillingness to eat genetically modified food because of concerns that it may be unsafe—or it may reflect the values of the group, values which may not necessarily reflect those of a wider public, such as opposition to cosmetic testing on animals. In both cases there may be subsequent social benefits, such as safer food and less abuse of animals. But pressure groups do not necessarily exist to reflect the interests of society at large, and social benefits are often indirect and sporadic.

Focus groups among shoppers who described themselves as "concerned about ethical issues," people who either had boycotted or would consider boycotting a shop or product because of ethical concerns, revealed that "only a small minority of ethical shoppers were purchasing ethical foodstuffs purely because of their ethical beliefs. A majority of respondents believed that these products were healthier and sometimes tasted better. . . . Ethical issues such as pollution of the environment, political oppression, or exploitation of the third world were 'back [of] mind' worries which rarely altered food-purchasing habits."[36]

In the case of foodstuffs the interests of environmental campaigners and individual shoppers often coincide, since both share an interest in safe, high-quality products. However, in many other cases, different consumer groups have divergent and conflicting interests, undermining their effectiveness as a bloc. According to Britain's National Consumer Council, "Consumers often find it hard to reconcile their concerns for the environment with their day-to-day needs as consumers."[37] In 1995, for example, Tesco bowed to consumer pressure by dropping cardboard milk cartons in favor of plastic bottles. The company claimed that research had shown a 92 percent preference for plastic. Yet the environmental costs associated with using plastic, which is energy intensive to make and more difficult to recycle, led to the move being criticized by green pressure groups.[38]

Shades of Gray

Perhaps the greatest flaws of consumer activism are its inability to deal with uncertainty or risk, and its need to reduce all argument to black and white, good and bad. Is GM food necessarily always bad for consumers or the environment? Or could this technology be harnessed for good, as Robert Shapiro, the chairman and CEO of Monsanto, initially seemed to have hoped, with his plans to eradicate famines in the developing world? Child labor may be distasteful to Western expectations, but does boycotting goods made with child labor improve or exacerbate the lot of third world children?

The inability of consumers—at least en masse—to see things in shades of gray is due partly to a media and NGO culture that thrives on headlines and soundbites; partly to instinctive Luddism (cars and frozen foods were both initially widely resisted), and partly to a general fear of change, be it beneficial or harmful. Monsanto now treads with such caution that when it was approached last year by Dr. Andrew Bamford, a genetic scientist at the University of Cambridge, to sponsor his work on using a rat gene to introduce iodine into rice, a development that could improve health in much of the third world, the company rejected his proposal for fear of consumer backlash. Consumer boycotts of goods produced by child labor may well succeed only in driving the practice underground and force vulnerable children into more degrading or dangerous work.[39,40] "Dismissing children (or not employing them in the first place) is tantamount to sentencing them to starvation."[41] And not only children, but their entire families, too: Children in third world countries are often their families' primary wage earners.

The world cannot be simplified to the extent that consumer politics tends to demand. Trusting the market to regulate may not ultimately be in our best interest. Such de facto populist politics can easily result in tyranny, not necessarily of the majority, but by those who can for whatever reason protest most effectively.

Rather than empowering all, consumer and shareholder activism empowers those with greater purchasing power and those with an ability to change their patterns of consumption with relative ease. It is a form of protest that favors the middle class—an expression of the dissatisfactions of the bourgeoisie. For the poor and socially excluded, those excluded from a wider range of goods and services by their low incomes and poor credit ratings, this form of protest is rarely an option.[42] Would the GM campaign in Britain have been so great if Prince Charles and the Women's Institute had not joined forces with Greenpeace and the Consumers Association?[43]

Governments Cannot Leave by the Back Door

In democratic states voting is premised on political equality: Each citizen has an equal opportunity to shape the political agenda. Consumer activism, however, is skewed toward those with superior resources and organizational abilities. If we neglect our rights as citizens we risk being marginalized as consumers. In turning away from traditional forms of expression, and embracing consumer direct action, protesters risk replacing representative democracy with a nonrepresentative alternative.

The emergence of consumer activism as a significant force clearly has political ramifications. Citizens can directly exercise power through bringing pressure to bear on corporations in a way that is increasingly denied to them through traditional political channels. Individual forms of protest such as complaint and litigation are increasingly replacing rather than complementing conventional forms of political expression. It is frequently the lack of trust in the honesty and competence of professional politicians that leads consumers to pursue direct action. The relative success of such protests, in comparison to the perceived ineffectiveness of politicians, further confirms the widespread belief that conventional politics is no longer relevant to most people's lives.

In a deregulated global market, where national governments are becoming spectators rather than actors, we see the consumer emerge as a powerful figure, bringing multinationals to heel through their decisions about where and how to shop. But just as consumer choice is premised on high-quality information, so also is it dependent on a framework of rights and regulations to protect the customer from unscrupulous vendors.

Because consumer power is market-based, it is effective only where consumers can convince a company that it is in its financial interest to comply. But consumer campaigns lack the legitimacy of democratically mandated protests, and so they are in this respect easier for powerful corporations to resist. Without the official weapons of sanctions, regulation, and restrictive laws, consumers are obliged to organize protests as best they can. Lacking the resources offered by the backing of democratic institutions, they can be ill-equipped to take on vested and powerful corporate interests.

Rather than providing an alternative to governmental action, the rise in consumer activism ironically makes it even more essential for governments to take an active role by providing the necessary information or by enforcing standards of transparency and accountability in business. Yet they have been extremely reluctant to do so. When the British government passed new legislation on consumer protection in 2000, the *Daily Telegraph* commented that it "places overwhelming emphasis on the power of informed customers to force down prices, while there are few solid plans for government action to force the pace."[44] The maxim of caveat emptor prevails.

But the Silent Takeover is not only about governance. It is not only about governments' unwillingness and inability to check corporate power. It is also about governments that are no longer able to deliver what their "customers" need. Is it possible that corporations will not only see the value in reacting to market pressure and deflecting criticism, but may also find a virtue in playing a proactively constructive role in society, too? Or that consumer politics is not only a stick, but a carrot that can lure corporations and business people into redefining their roles in society? Is it possible that while governments are allowing corporate interests to take precedence over those of the public, corporations and businesspeople may decide to put the public's needs first?

8

Evangelical Entrepreneurs

The Man Who Broke the Bank of England

Until Wednesday, September 16, 1992, few people outside the world of high finance had ever heard of George Soros, the Hungarian-American currency trader who had made a fortune in high-risk hedge funds on Wall Street. But on that day Soros, then sixty-two, made $1 billion profit, effectively by betting against the pound's ability to stay in the European Exchange Rate Mechanism. As a result Britain was forced to devalue the pound and crash spectacularly out of the ERM. Suddenly, Soros was a household name.

It was estimated at the time that Soros's winning gamble had cost every Briton twelve pounds in currency reserves lost by the Bank of England's increasingly desperate attempts to shore up sterling. With hindsight, most commentators believe Soros did Britain a favor by freeing it from the ERM, giving the country a chance to recover from the recession far more quickly than the rest of Europe. Either way, September 16—dubbed Black Wednesday—made both headlines and history, and Soros has subsequently been known as the Man Who Broke the Bank of England.

Yet his is by no means a straightforward story of the naked power of pure capitalism. Soros the financier is also one of the world's most generous philanthropists, a billionaire who in many respects simply wants to make the world a better place. Since the 1980s he has donated more than $1.5 billion to charitable causes, several times exceeding $350 million in a single year. And his high-risk approach to making money has spilled over into his charitable donations rather than into the concert halls and art galleries of traditional wealthy benefactors. Soros has consistently taken on high-minded, edgy projects in areas where governments fear to tread, from the legalization of drugs to the defense of Sarajevo against the Serbs. He has said, "When I was offered an honorary degree at Oxford, they asked me how I wanted to be described, and I said I would like to be called a financial, philanthropic, and philosophical speculator." Not surprisingly, his idiosyncratic approach to giving has attracted both admiration and disapproval.

Soros grew up in Hungary, the son of a Jewish lawyer who, by pretending that his family were Christians, managed to save them from the concentration camps. The young George, or Gyuri as he then was, was fifteen when the war ended. He subsequently lived under Communist rule in Hungary until, at seventeen, he traveled to London via Switzerland, where he worked in kitchens and studied at the London School of Economics. There he was influenced by Karl Popper, the philosopher who coined the term "open society"—a society in which argument and debate are encouraged, in other words, the opposite of a dictatorship. Soros has said that Popper's views have influenced him enormously throughout his career. In the late 1950s the young Soros moved into banking and from London to the United States, where he soon became an expert in arbitrage, the art of skimming off profits from buying and selling securities in different markets. He specialized in hedge funds, the high-risk investment vehicles which have the potential to make money whether securities or currencies rise or fall, by gambling on their future worth. Both he and his investors made huge profits. An investment of $100,000 in Soros's Quantum Fund in 1969 was worth $300 million by 1996.

But since the late 1970s his focus had been not simply on moneymaking. It was then that his first marriage broke up and he realized that he had neglected his relationships with his three children, something he later described as the biggest regret of his life. "I underwent a serious change in my personality during that period. There was a large element of guilt and shame in my emotional makeup." Soon afterward he started giving money away. In the 1980s he supplied the whole of Hungary with photocopiers, inspired by a desire to support the democracy movement in a direct way that would foster communication and make censorship difficult. He then approached both the U.S. and European governments with a seventy-year plan to assist the former Eastern bloc countries through the transition from communism to democracy. It fell on deaf ears. Frustrated, his stated aim in the Black Wednesday episode was to move himself from the City pages on to the front pages of the newspapers. "I had no platform," he explained at the time. "So I deliberately [did] the sterling thing to create a platform." Obviously people care about the man who made a lot of money.

Since then, governments have had to take Soros seriously. His initial charitable focus was on the former Eastern bloc. As a survivor of both Nazi and Communist regimes, he is passionate about supporting democracy in the former totalitarian states. He has spent $100 million to help post-Soviet science, given $50 million to support the besieged city of Sarajevo from Serb attacks, provided $13 million to aid projects in Belarus, plus millions more funding Open Society foundations (named after Popper's famous thesis) to finance educational and humanitarian projects around the world.

In the United States, however, he has attracted controversy for his opposition to what he calls "excessive individualism." In 1997 he warned that unfettered market capitalism could be as damaging to an open society as communism, prompting *Forbes* magazine to accuse him of talking "nonsense." His belief that drug addiction should be treated as a disease, not a crime, has led to him giving $1 million for a needle exchange project in the United States and being branded by the U.S. secretary of health as "the Daddy Warbucks of drug legalization." He has also attracted criticism for giving $50 million to help legal immigrants apply for full citizenship in the United States, and for sponsoring research on welfare and prison reform.

His touch on the markets has seemed a little less sure in recent years. His 1998 book, *The Crisis of Global Capitalism,* predicted a prolonged worldwide bear market following the Asian financial crisis, which, at the time, failed to materialize. He also failed to foresee the fall in technology stocks in the first months of 2000, with the result that his funds lost about a third of their value. But he is still branching further out into the worlds of politics and policy. He now runs his Open Society Institute from an office a couple of floors down from the Quantum Fund headquarters in New York, and he leaves most of the hands-on trading to his employees and his son Robert, devoting much of his time to philosophy and contemplation instead. He says, "I spend about a third of my day just thinking, and trying to clarify my own thinking about where I should be going, and where the world is going."

Despite all his philanthropy and recent misjudgment of the market, in 2000 the *Sunday Times* listed Soros's personal fortune at £2.4 billion. At least for the foreseeable future, Soros still clearly has the power to turn thought into action.

Political Philanthropy

George Soros is unusual but by no means unique. Many of today's corporate legends, presumably motivated by thoughts of the judgment of posterity, with their resources exceeding any imaginable desire for consumer goods and property and wanting to leave their mark, are deciding to assume a political and social role.

In a world in which governments seem increasingly unable to deliver, an increasing number of business leaders see it as their responsibility to do so. In the absence of forceful political leadership, many CEOs and company chairmen are choosing to involve themselves in wider debates. As economics has become the new politics, the tendency is for the purveyors of wealth to take over politicians' traditional roles. Why? It is not just another means of pursuing the bottom line; as Sir John Browne, CEO of BP, has said, "We are not put on this earth to facilitate easier driving to a video store."

In the competitive world of twenty-first-century business, mega-entrepreneurs also compete with one another to make the biggest social and political impact. As media billionaire Ted Turner told the American talk show host Larry King, "I'm putting every rich person in the world on notice. They're going to be hearing from me about giving more money away."

We are witnessing a new golden age of philanthropy, where the enormously wealthy are funding education programs and health initiatives, as well as more traditional donations to the arts and cultural institutions. While governments are facing stiff voter opposition to high personal tax rates and opposition from business to increases in corporate taxation, some of the rich are in effect taxing themselves. They are playing Robin Hood: taking from the state and giving to . . . well, whichever cause they deem worthy. A not insignificant number of people who have benefited so greatly from a system that is loaded in their favor are now turning to help those against whom the odds are stacked.

The founder of the Domino's pizza chain, Tom Monaghan, is devoting almost his entire $1 billion fortune to a nationwide school-building program in the United States, and has also paid for a hydroelectic dam in Honduras. Mike Milken, the symbol of 1980s financial lawlessness and amorality, now spends most of his time working to restructure society rather than companies. One of his foundations has awarded nearly $30 million to individual teachers, another funds cancer research. Peter Lampl, the British investment banker, is plowing £40 million of his money into schemes to help children from deprived backgrounds benefit from private education. But Bill Gates, the world's wealthiest businessman, a late convert to philanthropy, has trumped them all with his $21 billion foundation which will provide vaccines for the third world and education for its children. In real terms this is almost four times the total donation made by the hitherto greatest philanthropist, John D. Rockefeller, who started giving away his money when he was a clerk in Cleveland and by the end of his life had handed over $550 million, around $6 billion in today's terms, to various causes.

There are obvious parallels between nineteenth-century American robber barons like Rockefeller and Andrew Mellon and the twenty-first-century philanthropists: a desire to gild their reputations, to influence the future of their country, and, as Andrew Carnegie put it, to provide "ladders within reach upon which the aspiring can rise." But they differ in other respects. Their predecessors donated locally, today's philanthropists aim to make an impact all over the world. The robber barons had to be careful of their standing with government, their modern counterparts need to pay far less attention. Once, if powerful men overstepped the mark, governments eventually stepped in to curb their power, but today we see little evidence of that. The earlier entrepreneurs were rich, but today's are richer, owing partly to the significant rise in the real GNP of developed countries. Which makes the issue of what the superrich are to do with their money only more acute. Harvard Business School now offers a workshop in strategic philanthropy to address this very problem, and the most successful high-tech companies have full-time philanthropic counsellors to advise stock-rich employees.

Unelected Statesmen

Philanthropic funding is not only going the way of the arts and culture, museums and galleries, or charitable institutions. Today's philanthropists are much more political than their predecessors. With personal fortunes rivaling those of states, and a global presence that mocks states' limited reach, they are bypassing mainstream electoral politics to achieve political ends. Rather than seek election to office, many of them clearly believe that they can achieve much more as businesspeople than as politicians,[1] using the leverage of their business empires to gain access to world leaders and using that access to further their own diverse aspirations.

As Ted Turner has said, "I would only run for president if it was the only way I could get this country to turn around . . . I'm a deeper thinker. I've traveled all over. I have more access to information than anyone on the planet. When you realize your family, your friends, your society, and your planet is in a dire state of emergency, that has to change anyone with a responsible world outlook. I've thought about being president from time to time, and people have asked me about it from time to time, but I like my present job a lot more. I said back in the early 1980s that I want to be Jiminy Cricket for America. You know, the country's conscience."[2]

These real-life Citizen Kanes are effectively becoming a class of unelected politicians, ambassadors, and advocates, raising popular support, acting in defiance of government policy, donating money to supranational organizations, playing the role of unofficial diplomats, and using their power, wealth, and influence to effect political and social change to an unprecedented degree. Having made hundreds of millions or even billions of dollars in the corporate world, they now want to leave their mark in the public sphere. Peter Lampl has often said when questioned about his motivation, "What do you think I should do with my money, buy myself a plane?"

These new stars on the global stage, "chiefs" of "pan-global fiefdoms," are deferred to more than most world political leaders. Bill Gates had two summit meetings with President Jiang Zemin of China in eighteen months; Bill Clinton, by contrast, had only one.

Thanks to their wealth, acumen, and access, these businessmen are able to promote their beliefs and further their values. Beyond lobbying for changes that will directly benefit their business holdings, increasingly a significant number of leading businesspeople are lobbying for changes that will, in their opinion, better the world.

Take the case of the ice-cream maker Ben Cohen, cofounder of the Ben & Jerry's chain, who launched a grassroots campaign to derail plans for Hungary, Poland, and the Czech Republic to join NATO. He heads a business group, Business Leaders for Sensible Priorities; other members include Ted Turner, Paul Newman, and Alan Hassenfeld, chairman and CEO of Hasbro. The group broadcast a thirty-second commercial on several U.S. network talk shows in 1997, warning that NATO expansion would alienate Russia and rekindle cold war tensions. Cohen voiced concern about antagonizing Russia and the costs of NATO expansion. "Why would the United States expand a cold war alliance against a democratic Russia that wants to be part of western Europe when those resources could be used at home and abroad so much more productively?" he asked. "It's crazy."[3,4]

In January 1999 Business Leaders for Sensible Priorities began an advertising campaign opposing President Clinton's proposed $112 billion increase in defense spending. The United States, Cohen argued, is already spending over three times as much on defense as Russia, China, and "rogue" states such as Iraq, Iran, and North Korea combined. "Far and away, we already have the strongest military in the world. That's just not where we need to put our limited resources." He continued, "The real concerns we have as people are education and health care," and promised a "long-term campaign" to block the increase sought by Clinton.[5] While the American government was as usual lining up with the business interests of the arms manufacturers, a group of extremely powerful businessmen were declaring that this was not the best use of the state's resources.

Paul Fireman, Robert Haas, and Bruce Klatsky, CEOs of Reebok, Levi Strauss, and Philips Van Heusen, respectively, are also men with a mission. Recognizing that it is no longer enough for them to create "islands of sanity"—production facilities that uphold human rights—in countries where human rights abuses are the norm, the trio has crossed a line that most of their fellow CEOs and governments have until now been unwilling to cross. "I don't believe business should stand up and lecture governments on human rights," said Peter Sutherland, former director general of the WTO and now chairman of Goldman Sachs and cochairman of BP. Nor, as we have seen, do governments seem to see this as their responsibility in any effective or forceful way. But these three corporate activists are taking a stronger line.

In April 1999 they wrote to President Jiang Zemin, their goal being to persuade him to broaden union rights for the estimated four million workers who toil in China's forty-four thousand garment factories for as little as thirteen cents a day—at present only one powerless union is allowed. Levi Strauss even hired former assistant secretary of state for human rights Gare Smith as their point man. Over the next thirteen months they sent numerous letters and called several unconventional meetings with Chinese diplomats. At the time of writing, however, the meeting with Jiang has yet to take place. "We got the big brush-off," said one executive, summarizing a letter of March 10, 2000 from Beijing's ambassador in Washington. "He said, Jiang is busy for the rest of his life. Mind your own business."[6]

Their efforts give the lie to the claim—which, as we saw earlier, is used by many businessmen and politicians to justify continuing co-operation with repressive regimes—that the greater the access to foreign investment and influence the Chinese have, the more speedily their position on human rights will be reformed.

These CEOs are as yet unwilling to threaten pulling out of China, and the $500 million in annual exports and tens of thousands of jobs that they generate do not seem to carry sufficient weight with the Chinese regime. But at least they are nevertheless trying to use their insider status to good effect in raising with the Chinese government issues that politicians, afraid of jeopardizing trade interests, have in effect given up on years ago. A spokesman for the group said, "When you're doing international diplomacy on human rights with a country that hasn't respected them in a thousand years, just raising the issues must be viewed as a success."

But many of these unelected and self-appointed "politicians" are not just attempting to reshape the world, they are actually succeeding. The same traits that have enabled them to succeed in business—aggressiveness, self-confidence, acumen—coupled with their individual wealth, make them well-suited to pivotal roles in the political arena.

Ted Turner, for example, is not only an environmental crusader and social reformer who actively campaigns for cleaner transportation, wilderness conservation, and greener business; who over the past few years has given millions of dollars to environmental groups and has set up the Turner Endangered Species Fund to promote the conservation of U.S. species such as desert bighorn sheep, Mexican wolves, California condors, and black-tailed prairie dogs. He is also overtly political.

On September 18, 1997, he pledged to donate $1 billion to the United Nations, an amount roughly equivalent to the UN's annual budget and well over half the amount the U.S. government was in arrears at the time. Rather than donating money to a party in order to help his business interests, Turner has earmarked funds for his favored causes: the environment, children, population control, and women's projects. "My main concern is to be a benefit to the world . . . to control population, to stop the arms race, to preserve the environment," he said. "It's a billion-dollar day at the United Nations," announced a spokesman for UN Secretary General Kofi Annan, adding, "The moral boost it gives to the organization exceeds the financial value of Ted Turner's gift."

Businesspeople's involvement can be even more direct. Tiny Rowland, the buccaneering entrepreneur who once owned major mining interests in Africa as well as for a brief time *The Observer* newspaper, developed a plan with Scottish lawyers in 1997, which he subsequently helped to sell to Libyan leader Colonel Gadaffi, whereby Gadaffi would hand over the two men suspected of involvement in the bombing of the PanAm jet over Lockerbie in Scotland for trial in Holland under Scottish law.[7] Rowland clearly enjoyed outsmarting the British Foreign Office, but his involvement also consolidated his relationship with Gadaffi, an investment for the time when sanctions against Libya are lifted, as they undoubtedly will be.[8]

Boris Berezovsky, the Russian media mogul, has been involved in facilitating over fifty hostage releases in Chechnya, including the release in September 1998 of British aid workers Camilla Carr and Jon James after fourteen months in captivity. He has consistently denied paying cash for the hostages whose freedom he has secured but allegedly pays instead with the latest computer equipment. It is rumored that, thanks to Berezovsky, the residence of the Chechnyan rebel commander Salman Raduyev is now better equipped than some Russian special service units.

These and other business leaders share a desire to use their money and influence for political ends. They seem impatient with the speed of government action and have little compunction about bypassing traditional political channels; they dismiss the notion of state sovereignty and aim to override elected officials at will. Businesspeople are even involving themselves in peacemaking activities, a domain previously exclusively handled by diplomats and foreign ministers.

Monopoly Diplomacy

In 1993 Omar Salah, a twenty-five-year-old English-educated Jordanian, was driving down a California freeway when he heard the news that the Oslo Accord between Palestine and Israel representatives had been signed. Struck by inspiration, as he describes it, he flew to Israel to scout out potential business partners, where he met Dov Lautman, head of Delta Gallil, Israel's biggest textile company and the largest producer of private label underwear in the world, who was himself looking at the possibility of manufacturing in Jordan. There was instant chemistry. While the Jordanian–Israeli peace process took years to negotiate and sign, Salah and Lautman's joint venture, Century Wear, was in business eight months after the two men first met. Today it employs 2,200 people and is the largest private sector employer in Jordan, using Egyptian cotton, spun in Turkey, knitted and cut in Israel, and sewn in Jordan to make Calvin Klein underwear, Gap T-shirts, and Giorgio Armani boxer shorts that are sold in Europe and the USA.

Century Wear is only one of over forty alliances between Israeli and Jordanian businesses that have been set up over the past few years; over 50 percent of Jordan's leading businessmen are now involved in such ventures. Cooperation with Israel is essential to Jordan's business community, providing it with opportunities to acquire expertise, leapfrog technology, access otherwise elusive investment funds, penetrate global markets, enhance productivity, and provide employment.

The interlocking relationships created by these joint ventures are proving remarkably resilient. The business alliances have lasted, despite the antagonisms inherent in the whole Middle East peace process. Assassination and terrorism have been rife. Hundreds have been killed on both sides, a fact particularly poignant for Jordan, 70 percent of whose population is Palestinian. Yet not only have the alliances persisted, they have played a role in influencing the political relationships between the two countries. For now that they are Jordan's biggest private sector employers (with the joint ventures typically paying employees between 30 and 40 percent above local industry norm), no Jordanian premier can afford to jeopardize their future. At the Sharm el-Sheikh summit in October 2000 King Abdullah's response to the Palestinian uprising earlier in the year was surprisingly muted and he did not call for sanctions, evidence that these business relationships could potentially cement the peace in ways that politics could never achieve. After hundreds of years at war, it is businesspeople, rather than politicians, who are most likely to make peace between Jordan and Israel irreversible.[9]

It is not just in the Middle East that we can see the role that business can play in peacemaking. The Group of Seven, consisting of Northern Ireland's most significant trade and business organizations, created a second diplomatic track to the peace process when they held six meetings between October 1996 and the summer of 1998, in which they brought together representatives of all the political parties involved in the official peace talks and helped to mediate between the various factions. Shortly after the signing of the Good Friday Agreement, Sir George Quigley, chairman of Ulster Bank, one of Northern Ireland's largest financial institutions, and chair of the Group of Seven, claimed that its efforts had "made it less easy for the parties to simply walk away."[10] When asked why the CBI (the Confederation of British Industry) had decided to join the group, Nigel Smyth, director of the CBI in Northern Ireland, claimed that economic self-interest was the prime motivator, and he emphasized the opportunities for economic growth that the CBI felt peace would bring.

The sons of the most powerful men in China and Taiwan—Jiang Zemin, the Chinese president, and Y. C. Wang, Taiwan's biggest tycoon—have become joint venture business partners in a $6 billion project to build China's biggest computer chip plant. Indications are that this top-level alliance will have a significant influence on the resolution of the long dispute between Taiwan and China over the Taiwan Strait. One China watcher has commented, "If the sons can work things out, then maybe so can the fathers. At the very least, the sons can pass top-level messages back and forth should China and Taiwan start a new dialogue."

But, of course, businesspeople are not always driven by a desire to make peace. Civil wars in Angola and Sierra Leone have been fought on the back of diamond and oil interests for years. Black marketeers played a significant role in maintaining the Yugoslavian conflict. And Chase Manhattan's Emerging Markets Group's infamous memo on the need to "eliminate" the Zapatista rebels in Mexico was posted on the official Zapatista website. As easily as corporations and businesspeople can act as ambassadors of peace, they can also fuel conflict.

What is clear, however, is that when businesspeople see it as in their commercial interest to do so, they are able to build islands of cooperation amid seas of conflict, creating links based on shared economic interests that seem to be proving remarkably durable, while politicians' efforts at peacemaking seem increasingly ineffective. In fact, business is changing the political dynamic. The evangelical entrepreneurs are not simply enacting states' foreign policy; they are determining and enacting foreign policy themselves by their own actions.

This development is a fundamental characteristic of this latest phase of the Silent Takeover. As we have seen, governments are foundering and state revenues are increasingly constrained, while the rich are getting richer. And when they do, the egoistic thrust of capitalism seems in a number of cases to be overtaken by an altruistic force. For personal reasons, often independent of their business goals, some evangelical entrepreneurs are using their wealth, as George Soros has, as a platform from which they aim to change the world—distributing welfare, championing the environment, backing economic reforms, assuming a moral stance. Are these businesspeople an unrepresentative minority, the contemporary equivalents of nineteenth-century enlightened industrialists such as Joseph Rowntree, Robert Owen, or the Cadbury brothers? Or is the seeming broadening of motivations more fundamental? While the business of government seems more than ever to be business, is the business of business by contrast increasingly becoming that of government?

9

Mother Business

Corporate Carers

The AngloGold goldmine, Vaal River, South Africa, 4:30 A.M. The elevators that drop a mile below the earth's surface are packed with migrant workers who have come here from all over the country in search of work. Dark, hot, and dangerous with no way out when things go wrong. Rock falls and explosions are commonplace. Yet the risks of working below ground are arguably no worse than those above. Of the fifty thousand workers in this mine, 40 percent are HIV positive. And this is not unusual. Current estimates are that half of South Africa's miners are similarly infected.[1,2]

With many of the miners hundreds or thousands of miles away from home, housed in single-sex dormitories seventy to a room, the mines represent the honeypot toward which flock South Africa's second-class citizens: women. With their earning opportunities extremely limited, many with children to feed, prostitution in the shanty towns that encircle the mines is endemic. Ten rand with a condom, twenty rand without, the illusory closeness of "flesh on flesh" snatched in the toilets of the Saturday night mines bar is seen by many as worth the risk.

TB is also spreading, among miners whose immune systems have been destroyed by the HIV virus. In the mine hospital four men die as a result of AIDS-related opportunistic infections every week, and this figure omits those who have been sent back home first. The reaction of the South African government has been, to say the least, inadequate, despite the fact that mining for gold, coal, diamonds, and platinum remains the backbone of the South African economy. As Alan Whitehead, director of health economics and HIV/AIDS research at the University of Natal in Durban, puts it, "Our next lost generation will be children orphaned by AIDS, but what is the government doing? Nothing. They are not even thinking about it." In fact President Mbeki had, until very recently, denied that AIDS is caused by a virus, adopting the views of American biochemist Peter Duesberg, who is widely viewed by AIDS experts as at best misguided, at worst as a heretic.

While the South African government is largely failing to provide basic health education and care, AngloGold, Gold Fields, Iscor, and other privately held corporations are taking on these roles themselves. Companies, rather than the state, have set up clinics to tend to the dying; companies, rather than the state, are producing poster campaigns to explain the dangers of unsafe sex, are financing free condom-dispensing machines, are sponsoring and running AIDS education classes for junior managers, encouraging them to pass on the information to their workers, and are even providing anti-retroviral drugs to a handful of employees.

The companies' health activities are clearly not only motivated by a sense of moral purpose but, more centrally perhaps, by pragmatic business considerations. Each employee infected by HIV costs a mining company approximately $15,000 a year once AIDS develops.[3] Workers with AIDS are less productive, often absent from work, present a greater health care burden, and result in a higher turnover of personnel. Experienced specialist labor is not easy to replace. Morale is eroded among the survivors.

Unless a cure is found, which currently seems unlikely, 40 percent of Gold Fields mining company's fifty thousand miners will be dead within the next ten years; it is easier for the South African government in effect to ignore the problem than it is for the mining companies, for which the cost of inactivity is likely to prove prohibitively high.

Denationalizing Compassion

There are an increasing number of examples of corporations assuming responsibility on a much wider basis in this latest phase of the Silent Takeover. Many of the roles that governments are increasingly unable to play effectively, many of the responsibilities they are less and less able to meet, are now beginning to be met not only by individual businesspeople but by companies themselves. An increasing number of corporations—the same ones that lobby for favors; that irresistibly pressure governments; that can, in effect, refuse to pay increased taxes and angle for subsidies and privileges; even, in some cases, that in the past have shown little regard for the communities in which they operate—are nowadays showing signs of a caring face; and not simply as a reaction to consumer and shareholder pressure or government legislation.

This is not simply a matter of contracting out government services to the private sector—the collection of garbage, the provision of school meals, the delivery of welfare, even the running of prisons. In these cases government remains the client. It sets the terms of performance, determines the agenda, can replace the service provider if it fails to meet given criteria, and still, at least in theory, remains ultimately in control.

Rather, there is emerging a new dimension to corporate activity, one that puts corporations in the role of welfare providers and social engineers, environmentalists and mediators, in which corporations assume the traditional functions of the nation state. Business takes over the role of government.

Poor Communities—Bad Business

In parts of the third world, in countries in which the state is so moribund that it cannot deliver even the most fundamental of public goods such as education, basic health, roads, and infrastructure, corporations are deciding to meet the shortfall themselves.

In Nigeria, for example, Shell spent $52 million in 1999[4] on a social investment program, building schools, hospitals, roads, and bridges, supplying electricity and water to areas that the government effectively abandoned in the early 1980s. In fact the company now employs more development specialists than the government. "Things are back to front here," said Brian Anderson, who ran Shell's operations there in the mid-1990s. "The government's in the oil business and we are in local government."

Shell has learned, from experience, that it is not ultimately to its advantage to perpetuate the corrupt systems that have characterized Nigeria for most of the past thirty years. In the past, 70 percent or more of upstream revenues from its oil projects typically went in taxes and royalties to the government, and it has been claimed that Shell alone was responsible for three quarters of the Nigerian government's revenues and about a third of the country's GNP. Much of this money went into offshore bank accounts. Almost nothing was put into building Nigeria for Nigerians. Corruption was virtually universal. After Ken Saro-Wiwa's execution in 1995, protests increased, not least against the major multinationals, especially Shell in particular, which had been working with the corrupt regime.

The damage to its image in the first world has not been the only cost to the corporation. Resentment, civil unrest, and instability in Nigeria have also proved costly. Pipelines have been blown up, oil installations have been invaded, work equipment seized, and rigs blockaded.[5] In the autumn of 1998 the unrest cut Nigeria's daily oil production by a third.[6] In June 1999 fifty young Nigerian men invaded a Shell station and closed it for five days, costing the company $2.4 million. During 1999 alone Shell faced forty-five separate incidents of hostage taking, involving over two hundred staff.

By not facilitating trickle-down in the past, by being perceived as not building infrastructure, tackling inequalities of wealth, or respecting the needs of the communities in which it operates, Shell has paid a price. The company has slowly come to realize that it does not serve its own interests if it is seen as ignoring the way the proceeds of ventures remain undisbursed to citizens of the third world countries in which it operates. Instead, it is beginning to ensure that benefits actually reach the communities in which it operates and to meet at least some of the local basic needs. Where people cannot turn to their governments for support, Shell is showing itself prepared to set up in dialogue directly with the local people. Where the state is corroded or in collapse, men whose trade is oil exploration and drilling are now needing to act as diplomats, politicians, and mediators.

Unrest, instability, and poverty are not conducive to doing effective long-term business anywhere in the world. Jaime Augusto Zobel de Ayala II, president of the Ayala Corporation, one of the Philippines' largest conglomerates, said in a speech to Asian businesspeople in 1995: "We all pay for poverty and unemployment and illiteracy. If a large percentage of society falls into a disadvantaged class, investors will find it hard to source skilled and alert workers; manufacturers will have a limited market for their products; criminality will scare away foreign investments, and internal migrants to limited areas of opportunities will strain basic service and lead to urban blight. Under these conditions, no country can move forward economically and sustain development. . . . It therefore makes business sense for corporations to complement the efforts of government in contributing to social development."

In the West, too, the environment in which business operates does matter, and companies are beginning to recognize that poor communities make poor business. Yet Conservatives and New Labour, Republicans and New Democrats alike have made it clear that they are either not prepared or are unable to tackle the negative consequences of the system they continue to champion. "The state cannot solve all our economic and social problems. Nor should it try," said John Reid, then Labour's Scottish secretary (and now chairman of the party), in May 2000.

Given that the current consensus, from both right and left, is that the answer to poverty is not welfare payments and government handouts—for governments are not prepared to fund social programs by running up deficits—many corporations are increasingly seeing it as good business practice to take on the role of societal custodian themselves. Sir John Browne, chairman of BP, said, "The simple fact is that business needs sustainable societies in order to protect its own sustainability."[7]

Companies are beginning to realize that social and environmental issues cannot just be tacked on to their overall strategy as insignificant addenda; they bear directly on company performance. In the short term, corporate interests can ignore the wider interests of society, but if society breaks down altogether, corporations also have to bear the cost.

Believing that companies cannot remain aloof and prosperous while the surrounding communities decline and decay, Ben Cohen and Jerry Greenfield made it a condition of the sale of their ice cream business to Unilever in April 2000 that Unilever create a $5 million fund to help minority-owned businesses and others in poor neighborhoods. Rite Aid, the American drugstore chain, has committed 580 of its 4,000 stores across America to inner-city areas that are in need of revitalization, contracting with local developers, hiring local employees, and creating positive ripple effects in the local economy. There is no reason to imagine it is doing this only for compassionate reasons, however; it makes good business sense, too. CEO Martin Grass acknowledges that inner-city involvement can be attractive to the company because the disposable income of such areas has been undervalued in the past. There are 7.7 million households in depressed inner-city areas of the United States, spending more than $85 billion per year on retail items, even though many are without access to local quality goods and services;[8] inner-city areas thus present a huge and unexploited commercial opportunity.[9] And also a huge unexploited political opportunity for politicians, for this same market also consists of areas with the greatest latent and untapped voting power.[10]

The Corporate Nation

Government's retreat has presented business with an opportunity. As politics becomes an ever more fuzzily branded product, corporations can both take the moral high ground and gain real business advantage by assuming social and environmental responsibilities.

Schools have become a prominent battleground for corporate marketers who want to maximize the potential value of social contributions. While governments, in today's environment of low taxes and redirected priorities, are ever more pressed to find sufficient resources for education, companies are seizing on new business opportunities. Go into any classroom now, and the quantity of products "donated" by corporations is startling. In countless British classrooms the Tesco computer stands next to one bought with vouchers from packets of Golden Wonder potato chips.[11] United Biscuits has distributed to eight hundred British playgroups and primary schools a kit that encourages children to learn their reading and arithmetic by discussing "their favorite Penguin biscuit" and copying the "P-p-pick up a Penguin" slogan. In the United States, McDonald's has supported literacy programs by attaching booklets with tips for reading aloud to 13 million Happy Meal bags. Nike is among the biggest contributors to parks, recreational facilities, and other youth-related projects. "Give me a child before the age of seven . . . ," said the Jesuit priest; companies are taking this maxim literally, some by producing grab bags of branded items for nursery school-age children.

Procter & Gamble, Toyota, and Marks & Spencer are just a few of the many companies that are providing advice, funding, and managerial support to help children to develop basic skills and education. Levi Strauss runs schemes aimed at young truants, seeking to provide them with routes back into the school system. Accountancy firm KPMG runs a teacher mentoring system in the U.K. and South Africa. BP's "Science Across the World" initiative operates in thirty-seven countries on six continents, reaching sixty-five thousand students. Bill Gates has announced his intention to put a computer on every British child's desk. And in 1993 Honda spent $17 million to set up a remarkable school in Colorado for students whom the public school system is failing, providing residential care for eighty high school seniors from all over the United States in a 640-acre setting on the eastern slopes of the Rocky Mountains.

Here wayward kids, children who have skirted the mainstream of society, persistent truants, many with problems with drinking or drugs, teens with nose rings, mohawks, and platform shoes, benefit from Honda management's mandate to make a social impact in ways that will not be perceived as too self-serving, in ways "that go beyond what one would expect of a corporation."[12] The school is a resortlike campus that not only has six housing units but also a full-length soccer field, three classroom buildings, and a gym complete with weight room, basketball court, swimming pool, and climbing wall. An all-purpose lodge where pupils watch TV, eat, and attend daily morning gatherings at which they confess their misdemeanors, hug, and make clear when things are outside of their "comfort zones aids their social development." The school doctrine is "tough love," and students soon develop a firm sense of right and wrong. Over half of the school's graduates have gone on to college. All but a few of the rest are either working or in the military.

Honda contrives to support the school to the tune of $3.5 million a year, which guarantees free year-round tuition for the students. This works out at roughly $25,000 per student each year, a figure several times the average in U.S. public schools. The children are provided with all their essential needs from books to gym clothes, as well as some nonessentials such as long-distance phone time. And the school has become a model for other alternative schools elsewhere. Each year as many as two thousand teachers, principals, and scholars pass through the campus, sitting in on classes and meeting with students, staying in a bunkhouse specially built to accommodate them.[13]

Other corporations are working to meet other social needs. By investing in such nonbusiness-related areas, by addressing the downsides of capitalism and globalization, companies aim to distinguish themselves as caring about more than simply making another sale. By investing in the community, they can improve their images, lower employee turnover, and increase their chances of getting the people's "vote."

Entrepreneur Anita Roddick's entire business is focused on the virtues of ethical behavior and social engagement. Her Body Shop brand has become synonymous with social activism, support for human rights, and the protection of the environment and animals, all through selling toiletries and cosmetics.

BT, the telecommunications company, published a paper in 1998 entitled *Changing Values*. The values referred to were not shareholder value, or the value of the brand, but those concerning world poverty and the widening gap between rich and poor.[14]

Procter & Gamble tackles education, crime prevention, and economic regeneration and is using its skills in marketing, communication, project management, and staff mentoring to contribute to the communities in which it operates. In parts of Newcastle in the U.K., for example, in which three quarters of the children live in households with no earned income, P&G has developed a mentoring and business skills program for local schools and used its marketing expertise in efforts to combat drug abuse in the area.

Avon, the American beauty products company, has tackled health issues and launched the Breast Cancer Awareness crusade, which raises money for breast cancer programs through sales of Avon "pink ribbon" products and lobbies for more government-funded research.

And some of the biggest names in corporate America, including Microsoft, PepsiCo, Coca-Cola, Boeing, Kellogg, General Motors, Procter & Gamble, and Bank One, went on a collision course with President Bush in early 2003 over the issue of affirmative action in U.S. universities, putting their signatures onto a brief filed with the U.S. Supreme Court arguing that universities should be allowed to consider race as a factor in admissions. This is a position criticized by George W. Bush.

What motivates these companies? Why is Honda, a Japanese car corporation, funding a school in Colorado? Procter & Gamble tackling drug abuse in Newcastle? PepsiCo and General Motors taking a stance on the issue of affirmative action in U.S. universities?

In part because employees and managers do not leave their values as individuals at home when they arrive at work. Now that belief in the integrity of political parties is at an all-time low, employers can foster staff loyalty by espousing employees' values. And those employees who are in demand, those with highly developed and specialist skills, are becoming increasingly selective about who they choose to work for. While the choice of political parties is in effect highly restricted, this elite can choose to join organizations whose values they find sympathetic.

M.B.A. graduates nowadays, for example, increasingly want to find employers whose values reflect their own views on social responsibility and corporate citizenship and place a strong emphasis on the importance of the environmental and social reputation of a company when considering it as an employer.[15] Several multinational oil company managers are finding that during job interviews many of the most promising university graduate engineers are now asking about policy on the environment and human rights, questions which were rarely raised in the past. Numerous studies conducted in recent years[16] have shown that "employees prefer to work for a company that has a good reputation as a corporate citizen," and that allegations of corporate irresponsibility have a serious effect on morale.[17] Directors of both BP and Shell, for example, report being overwhelmed by the large number of concerned e-mails they received from staff following adverse TV and press coverage[18] of their operations in Colombia and Nigeria. And in the wake of the September 11 attacks, four out of five Americans said that they believed that it was more important than ever for their employers to support the needs of society. Three quarters of those questioned said that a company's commitment to social issues is an important factor in their choice of who to work for—up from 48 percent in March 2001.[19] It seems that faced with a recession and feeling vulnerable due to external threats, the American worker is turning to his employer to provide him with some sense of moral certainty. As *Fortune* magazine has noted, the single most reliable predictor of overall excellence in a company is its ability to attract and retain talented employees, and companies are listening.

Companies have also realized that by playing a greater role in the welfare of the communities in which they are active, they can improve the bottom line: 86 percent of British consumers say they have a more positive image of a company if they see it is "doing something to make the world a better place."[20] A third of American consumers said that the primary goal of business should be building a better society[21]—and are willing to support those that do, seizing the chance to make a positive contribution through their purchasing decisions. Coca-Cola calculated that in 1997 it experienced a 490 percent increase in sales of its products at 450 Wal-Mart stores during a six-week campaign allied with Mothers Against Drunk Driving, to whom the company donated a proportion of its sales. Diageo PLC, the food and drink company whose brands include Johnnie Walker and Smirnoff, reported that between 1994 and 1998 twenty-two cause-related marketing projects helped it raise $600,000 for causes while increasing the sales of tracked brands by 37 percent. Wendy's International in Denver increased sales of jumbo fries by more than a third in 1998 when it contributed a percentage of each purchase to the city's Mercy Medical Center. Unlike the payment of taxes, whereby people have little sense of where their money is being spent, cause-related purchases enable them to earmark where their spending is going. Consumers will change brands, switch retailers, be more accepting of price rises, and have a better perception of a company when businesses or brands are linked to a good cause. Cause-related marketing enhances corporate image, builds brands, generates PR, and increases sales. No wonder the linking of sales to good causes has increased threefold over the past decade, and over 85 percent of American corporations now use cause marketing.[22]

Sheep in Wolves' Clothing

In the world of the Silent Takeover corporations are beginning to realize that customers will reward them, not only for not doing ill, but also for being seen to do good. Consumers and workers "no longer buy the product, they buy the entire company ethos. What companies do, make, or sell is inseparable from what they are." Corporations are discovering that by taking on some of the costs that governments have relinquished, they can improve their standing in society and consequently their profits. Does this represent a win-win solution?

It is precisely with regard to those issues that the state finds hardest to regulate, issues that transcend international borders and that could be addressed by politicians only through hard-to-enforce international protocols and international agreements, that we see the inherent advantage that corporations have in tackling world issues. Operating globally, companies can assume responsibilities, make decisions, and stimulate concern in ways that governments find extremely hard to attain, and at a speed that government bureaucracies are unable to equal. Companies can even take the initiative. As it says on the BP website, "It is not enough . . . to rely on the leadership of politicians. Business needs to play a responsible, active role and to show leadership in finding solutions and putting them into practice."

So while the U.S. government has still not ratified the Kyoto Protocol on climate change, which was agreed on in principle back in 1997, a number of oil and chemical companies are already putting its aims into practice. Lord John Browne, CEO of BP and the first oilman to declare that global warming may be a real issue, has led the way with his pledge that BP will decrease emissions of the greenhouse gas carbon dioxide by 10 percent from 1990 levels by 2010, which is significantly more than both the United States's promise of 7 percent and even the EU's more aggressive 8 percent; and also with his implementation of an internal emissions trading system whereby BP divisions can trade pollution credits with each other, a system discussed at Kyoto, but still not implemented at any national level. Other companies, fearful of being singled out as environmentally unsound, are following suit—with Shell, Dupont, Alcan, Pechiney, and others agreeing to match BP's targets, and either also trading pollution permits between their subsidiaries or developing trading systems themselves. And one of the more remarkable sights at the Johannesburg Earth Summit in 2002 was the representatives of big oil firms, nuclear processing companies, chemical conglomerates, and car giants making alliance with their historic foes in Greenpeace to issue a joint plea for governments to meet the Kyoto targets for reducing greenhouse gases. It was, until recently, utterly unthinkable that the eco-warriors and the multinationals would find common cause. As Bjorn Stigson, a vice president of the World Business Council on Sustainable Development, said: "This is a good example of where the need to save the planet is so important it transcends any other differences we may have. We will no doubt disagree about much else at this summit, but we are united in the belief that we must stop climate change before it is too late."

Who Is Taking Over Whom?

While politicians shirk their social and environmental responsibilities, pleading the pressures or limitations of globalization and economic interests, corporations are, as we have seen, increasingly taking over this mandate in directly tackling the world's global issues. For the opportunities of the global market have come at a price. The greater a corporation's reach, the more visible it is, and the greater the benefits of being seen to contribute to the remedying of the world's ills.

Many of our leading companies now realize that responsible behavior and social engagement are increasingly less of an option, and more a necessity, if a company means to safeguard its commercial interests. Big business has begun to move beyond seeing social responsibility as an effective PR tool; corporations are now being held hostage by their own marketing strategies. Not least, in the need to compete with each other, the causes they champion must be ever worthier, ever more substantial, and ever more original. The innovative campaigns of the Body Shop during the 1980s have become the humdrum promotions of today's main street banks. Not only do products have to be seen to be whiter than white; company ethics, ideologies, and initiatives have to be so, too.

So we see the spring shoots of the latest phase of the Silent Takeover: corporations taking stands against incumbent governments, actively pushing forward political agendas, substituting for the policy failures of politicians. Corporations are beginning to right social ills and dispense justice in places where the state does not, to tackle issues of environmental degradation while governments seem paralyzed, to champion human rights in those parts of the world that no politician seems prepared to speak out for. Having bankrolled a system that creates losers as well as winners, corporations rather than governments increasingly seem to be the only institutions that are attempting to address the imbalance with any effect.

Politicians are actively supporting this unexpected development. In February 2000 Bill Clinton, then president of the richest and most powerful nation, addressed a group of the world's most influential business leaders in Davos in Switzerland. "My most important wish," he told them, "is that the global business community could adopt a shared vision for the next ten to twenty years about what you want the world to look like, and then go about trying to create it . . . collectively you can change the world."

The truth of the matter is that businesspeople are often able to do more than governments can. They are not tied up with bureaucracy; they can bypass protocol, make decisions single-handedly, and have a direct impact. They are able to transcend national boundaries and to ignore the strictures of supranational organizations. They often have access where foreign ministries and diplomats cannot reach. While the British government was powerless when the human rights activist James Mawdsley was jailed in Burma for seventeen years for distributing leaflets criticizing the country's military junta, it was

Premier Oil, the British oil company vilified for its support of the Burmese regime, that was instrumental in securing his release. As Charles Jameson, its chief executive said, "We are in a unique position. . . . We are the only British organization which has access at all levels."[23]

It is in fact a double switch: Politics has entered commerce, consumerism has entered politics. Politicians, by not providing the same levels of service in hospitals and schools, the same quick response to our concerns, and the same willingness to tackle the downside of the laissez-faire system as do P&G, BP, and Honda, are forfeiting our custom. Corporations, realizing how easily transferable our "vote" is, are becoming more responsible and responsive in fear of our possible "defection." We don't have to wait four or five years for an election, after all, to change our product choice.

But who, in this latest phase of the takeover, is taking over whom? Politicians are spending some of their time acting like salesmen, and corporations some of their time acting like politicians. Consumers are voting with their pockets while the electorate is increasingly staying away from the polling station. Governments are emulating business and business is emulating government. The state is giving up many of its responsibilities, and corporations are beginning to take them over in its place.

Since the eighteenth century Western societies have regarded politics as essential to the progressive enactment and widening of reforms. Today we are seeing that other institutions are able to take on some of these roles. As our expectations of politicians are declining, our expectations of corporations are becoming both greater and broader.

Does this mean that the companies that succeed in the twenty-first century will be those that decide that the business of business cannot solely be business? Is it just possible that corporations, whom we are used to expecting to disregard human interests in their pursuit of profit, are in fact turning out to be our reliable allies? Are their attempts to right injustice constructive? Are their remedies more effective than those of government? Is their willingness to effect our will greater than that of our elected representatives?

Can this fusion of consumer politics and corporate power provide satisfactory solutions to the problems created and encouraged by an untrammeled capitalism, or even be a satisfactory replacement for traditional politics? Or is it a chimera? And if it is a monster, will it devour us?

10

Who Will Guard the Guards?

Soylent Green

It is 2022, and the Earth's face has changed. The greenhouse effect has raised average temperatures significantly. In most cities the polluted smoggy air is virtually unbreathable. New York has become an overcrowded hellhole and now has 40 million to feed.

The rich are richer and the poor poorer than ever—sleeping wherever there's room and fighting each other for scraps of food. For the poor, most of life's simplest pleasures are a thing of the past. The rich live in separate luxury apartments in gated communities and have access to such luxuries as paper and pencils, bars of soap, and running water. The only time people see blue skies and green forests is on video at the government-run euthanasia centers during their last twenty minutes of life.

Natural foods, fruit, vegetables, and meat are almost extinct. Strawberries cost $150 a jar. The only food to which the starving masses have access is that manufactured by the Soylent Corporation, a powerful food-processing company—concentrated nutritional wafers that come in red, yellow, or the more nutritious green produced, according to the TV commercials, "from the finest undersea growth." But even this is in such short supply that when it is distributed riot police need to be available to try to limit the resulting violence.

In the classic 1973 sci-fi movie Charlton Heston plays Detective Thorn, a man who has known nothing but this dystopian world of overpopulation and global warming, and whose mission is to investigate the murder of an official from the Soylent Corporation. But in the course of his investigation, he discovers a much bigger crime. The executive, it is revealed, was killed because he knew something terrible: the secret ingredient of Soylent Green. For in order to meet the demand for their product, the powers that be have secretly taken to making these wafers not out of soybeans and plankton as advertised, but out of the many newly dead.

"You've got to warn everyone and tell them!" says Thorn in the final speech of the film. "Soylent Green is made of people! You've got to tell them! Soylent Green is people!"

As the year 2022 approaches, these are essentially the issues we face. Is the world we are heading toward, this emerging world of "evangelical entrepreneurs," "corporate carers," and "campaigning corporations," a world in which corporations really will safeguard the public interest? Or are we are heading toward the apocalyptic world of *Soylent Green*? A world in which the corporate interests—literally or metaphorically—feed off our carcasses? Will the corporate takeover ultimately be for good or ill?

Corporate Aid

In the third world, in places where the state has collapsed or is greatly weakened, the corporate contribution to welfare—its assumption of quasi-state roles, its ability to ensure that rights are upheld in places where the political will to do so does not prevail, its ability to mediate between warring factions—can provide a lifeline to people abandoned by their own governments.

Is it ideal that corporations, rather than host governments, perform this role? No, of course not. Corporations, and their managements, are unelected. These functions are alien to their core business—managers of multinationals operating in the third world are often overwhelmed by the social problems they encounter, and understandably find it difficult to know which causes to prioritize. They have no expertise in distributing aid, and limited experience in the provision of state functions, even though they are often working closely with NGOs and grassroots organizations, and hiring knowledgeable people. Their contributions can be squandered or diverted through corruption. IBM provided a number of computers to South African schools, but these schools were highly prone to theft. In schools computers are only useful if they are accompanied by appropriate teachers, not, of course, provided by IBM. So the impact of the contribution was minimal, a significant number of children did not become computer literate, and money spent on welfare was, as is often also the case in the public sphere, completely wasted.

There is a real risk that as and when corporations move on (we have seen how easily modern companies can change the location of their activities), so, too, will their obligations to local communities. There is a real danger that their presence, their taking on of these traditional governmental roles, will create a disincentive for governments to develop appropriate institutions themselves; and that if and when they pull out, there will be nothing at all to replace them, and no recourse to be had. There is a real danger that corporations could use this dependence to exact a stream of IOUs and quid pro quos, to demand ever more favorable terms and concessions from host governments. There is a real danger that such "socially responsible" policies could be a form of "greenwashing"—where, hiding behind a superficially responsible screen, corporations can abuse their considerable power and inflict damage on the society or the environment in which they operate.

All these concerns are valid, and they should not be dismissed. The history of multinationals overseas has often been shameful: more often rapacious than peace-seeking or just; characteristically silent when faced with regimes that violate civil and political rights. And there is evidence that current practice may not always be that different. Coca-Cola is currently being sued in the American courts for what the lawsuit describes as "the systematic intimidation, kidnapping, detention, and murder" of workers in Colombian bottling plants.[1] However, the amount of money Western governments are willing to provide for overseas aid does not begin to address the needs of developing countries, and many recipient governments seem unwilling to put their people's needs first, either by distributing the gains of trade among them or by enforcing regulation. In many places where multinational corporations operate, the state lacks legitimacy and is undemocratic. We cannot afford to dismiss the fact that business is now in many ways better placed than any other institution to act as the primary agent of justice in much of the developing world.[2]

In the West, also, it is again business that is increasingly appearing (and with governmental encouragement being put forward) as a potential if unlikely savior for at least some of those whom untrammeled capitalism has failed. These developments are in their early stages. But as governments operating in global markets find it increasingly hard to raise the finances to fund social welfare programs at appropriate levels, particularly in the context of changing populations, and the inevitable downturn in the business cycle causes the government surpluses of today to turn to deficits, we will in all likelihood see more welfare functions being provided by companies, both as stand-alone for-profit business ventures and as add-on features to existing product ranges.

In some respects this is an attractive development. For not only has the private sector's delivery of public services often proved superior to public provision but, given the interminable bureaucracy inseparable from traditional politics, government's loss of its monopoly on politics is probably ultimately to our advantage, should these developments be managed correctly.

Who Will Guard the Guards?

But that is the crux—"managed correctly." For the more we come to depend upon corporations in areas crucial to our basic welfare, the more critical it is that if things go wrong, they can be held accountable. But if corporations take over the role of governments, "who," as Juvenal first asked in A.D. 100, "will guard the guards themselves?" How will we ensure that Soylent Green is not made of people? We have already seen how governments' ability to check corporate power is under threat from supranational organizations such as the WTO. Will governments be able to be effective regulators if they become ever more dependent on corporations to deliver welfare solutions? Will companies not be able to use positive activities in one context to protect themselves from being held accountable for the harm they may be doing elsewhere?

In the Hungarian town of Pecs, where it has a large cigarette factory, British American Tobacco (BAT) has sponsored a medical clinic, a hostel for the homeless (now named BAT House), and a theater (the BAT Theatre). While restrictions not only on cigarette advertising but also on commercial sponsorship by tobacco companies seem likely given Hungary's forthcoming ascension to the EU, in a country in which the state is unable to provide adequate welfare provisions, a tobacco company can market its brands not only by associating with good causes but can also use this to shape their political relationships. BAT company executives now sit on several local government committees and, as Pecs Mayor Lazlo Toller has said, are invited to all "high-level meetings."[3]

Ted Turner's donation to the UN, noted in Chapter 8, raises similar concerns. For although his intentions are undoubtedly philanthropic, his donation sets a dangerous precedent. With America's payments to the UN in arrears of more than $250 million, could the UN be forced to turn to wealthy businesspeople or corporations to fund itself, becoming "a club for wealthy individuals[?] . . . scary, given that not every wealthy donor's agenda may be as benign as Turner's."[4]

This concern was first raised in 1958 when the U.K. floated a proposal—immediately quashed—to allow the UN to raise money from philanthropic businesses. "With money inherently comes ties," explains the historian Rachel Thourt of the United Nations Research Group. "The smaller nations asked a good question: If the UN takes money from oil companies year after year, it will grow dependent on it. Even if Standard Oil says there are no conditions, won't we have some desire to keep them happy? Then won't it be harder to sort out things in the Middle East?"[5]

As we have seen in previous chapters, the market cannot be counted upon to ensure that corporations will always act in our best interest, so we must be able to continue to count on government's ability to play the role of regulator of last resort. Regulation with teeth is clearly often needed to counter monopoly power, safeguard the rights of individuals, and protect the community from the abuse of corporate power more generally. But if government and business become "partners," who will be there to adjudicate if things go wrong? In the early twentieth century the American government passed strong and effective antitrust laws to protect the interests of society when business got too big and powerful. The growing dependency of government on business, and the blurring of boundaries between the public and private sector, compel us to ask who will step in now.

This latest phase of the takeover raises other questions. For example, do we want unelected businesspeople and corporations stamping their opinions and views indelibly on our own lives, however seemingly benign their intent? Do we want to risk our world becoming one in which we have to depend on corporate charity?[6]

Politicians talk of "market-based solutions," "private delivery of public goods," and "corporate citizenry," as if these were the answers to the failings of laissez-faire capitalism. But it was the American government, not business, that picked up the pieces after the crash and depression of the 1920s; and government money, not corporations', that rebuilt European economies after World War II and created welfare states.[7] Corporations may be able to play some role in alleviating poverty and tackling conflict and inequity, but social investment and social justice will never become their core activity. Their contribution to society's overall needs will always remain at the margins, and their contribution to welfare will never be comprehensive.

Unlike politicians, who are charged with looking after their citizens' interests, businesspeople and businesses have no such mandate. Even where wealthy individuals have other motives, those of companies will always be predominantly commercial rather than moral and so will be subject to market vagaries. For it is, of course, those social concerns deemed most appealing to their customers that will be embraced by corporations in the West. It is charity predicated not on need but on market appeal. And in a world where welfare and social justice are increasingly left to the market, minority interests or unattractive causes may well suffer. The sick, the homeless, those of us with limited purchasing power potential, or a lack of customer appeal—those with little voice—risk being excluded even more than now.

Despite the inefficiencies of public service provision, despite the fact that many goods and services are better delivered by the private sector, when public services are taken over by the private sector, profit becomes their raison d'être. If the government does not manage these private providers of our needs stringently, and admit that not all public goods should be for sale, tragedies like the millions of uninsured infirm of America, the Paddington and Hatfield rail crashes in the U.K., which occurred as a direct result of the way in which British Rail was privatized, and the airport security lapses in the United States on September 11 could become more commonplace.

Even before September 11, investigators for the General Accounting Office, the federal watchdog agency, had issued very critical reports about the failure of airport security, owing to the uniquely American system of letting airlines pay for and be responsible for security. "It's considered a way of life for the airlines," said Mary Schiavo, former inspector general for the Department of Transportation. "They get fined and consider it a cost of doing business." The arrangement was that the commercial airlines hired private contractors to provide security. Typically the contractors they hired were those who had presented the lowest bids. These contractors would go to the lowest-cost people out there—minimum wage people—and provide them with as little as eight classroom hours and forty hours of on-the-job training, to save costs. And this cost-saving strategy persisted even after the terrorist attacks. Argenbright Security, the country's largest airport screening contractor, was discovered by the Department of Transportation and the Federal Aviation Administration to be in violation of federal regulations at 124 airports, including the employment of illegal immigrants after September 11. Whatever the arguments in favor of turning over social responsibility to the private sector, the parallels with the privatization of public services provide significant cause for concern.

When the Party's Over

Moreover, corporations were able to build a business case for social responsibility during a period of unprecedented economic boom. Their customers, most of whom were reaping the rewards of economic good times, proved to be—and were able to afford to be—socially minded. They rewarded companies for taking on welfare roles, and in some cases even showed that they were willing to pay more to have companies dispense global justice. But will customers' priorities change if the economic downturn persists and, as a result, corporations' priorities, too? Will the business case for social responsibility be sustainable during this ongoing recession? When cost-cutting becomes essential for corporate survival, when the emphasis on price will again be paramount?

Until recently in Japan, trading groups, *keiratsu,* once provided extensive social security systems for communities. It used to be the norm that companies would spend up to 70 percent on top of actual wages to ensure the provision of welfare systems—"corporate communities," they used to be called. And many of the welfare functions normally associated with the state—housing, job creation, local economic development, education—were provided by corporations. But in the wake of the Asian financial crisis and faced with the demands of ever more competitive global markets, Japanese companies have been unable to continue these practices. The lifetime employment system that effectively provided social security for the nation has disappeared and with it the security of countless families. The head of Toshiba says that they are no longer "a charity," corporations have been selling off their company dormitories, and bonuses have been reduced. Entire communities are suffering in the Nissan towns, the Mitsubishi villages, and the Toyota cities which have depended on the social network and security provided by the *keiratsu* system for nearly fifty years, as plants close down and firms withdraw support from the community. School vouchers, health care, and many other provisions are being reduced or withdrawn entirely. "The suicide rate in Japan was up one third in 1999 over 1997, a testament to the social strain."[8]

The Japanese example provides a warning to those of us who have looked to the corporate takeover of welfare as a possible solution to the social problems created by laissez-faire capitalism. For if this move toward greater responsibility and care that we are beginning to see in the West is predicated solely upon the continuing strength of the global economy, upon the fact that philanthropic acts are essentially tax write-offs against balance sheets firmly in the black, it is surely liable to be reversed when times once again become difficult. When in 1999 Nestlé withdrew from sponsoring London's Notting Hill carnival at the last minute, claiming that its new cold coffee drink (which was to have featured prominently) was not yet in the shops, what hope is there that in a downtown commitments will be upheld?[9] Unless the impact on their reputation of withdrawing from social commitments is deemed more costly than that of maintaining such commitments, companies will simply not be able to justify staying involved to their shareholders. The corporate provision of welfare risks dependence on the continued generation of profits.

And the Future?

Despite the fact that major companies will increasingly do their business in the context of potentially critical public opinion, there remains great uncertainty about the future. What happens at the point where ethical business considerations and profit-making diverge sharply? If companies decline to pay the increased costs of social responsibility, preferring to relocate or withdraw their investment, it is not clear that either politicians, pressure groups, or individual consumers wield sufficient power on their own to prevent them.

Is there a price that will be exacted for acts of corporate benevolence? Today Microsoft puts computers in our schools; will it tomorrow determine what our children learn?[10] When Mike Cameron, a nineteen-year-old student, turned up at Greenbriar High School in Evans, Georgia, on an official "Coke Day" wearing a T-shirt with a Pepsi logo, he was suspended by the school authorities. "I know it sounds bad," said school principal Gloria Hamilton. "It really would have been acceptable . . . if it had just been in-house, but we had the regional president here and people flew in from Atlanta to do us the honor of being resource speakers, and Coca-Cola does so many positive things for our school like helping organization and sports. These students knew we had guests."[11]

Tesco's Computers for Schools campaign, sponsored by 7-Up, Tango, and Pepsi, may get computers into schools, but it also increases children's exposure to sugary drinks that can damage their teeth. Vending machines are generating thousands of pounds per year for some secondary schools, income that is crucial given problems of inadequate funding and lack of resources. "It's hard for head teachers to resist," says Joe Harvey, director of Health Education Trust, a U.K. charity dedicated to food policy development in schools. "You suddenly think, Great, now I can afford the part-time classroom assistant that I need for the special needs department."[12] Yet only one in ten products targeted at children was found by the Food Commission to be healthy.[13]

The American Channel 1 network is by now notorious for having provided twelve thousand American schools with money and supplies in exchange for being able to broadcast commercials into classrooms.[14] In Germany a number of schools have responded to declining government funding by turning to corporate sponsorship, with Coca-Cola, L'Oréal, and Columbia Tristar only three of a number of companies now allowed to put up posters in German schools. And in 2001 above the Brooklyn Bridge there floated a painting of a Snapple bottle, created by a nonprofit arts group, and sponsored by—you guessed it—Snapple. But do we want to live in a world in which commercialism takes advantage of shortages in funding and rides off the back of children's learning? A world in which this could be the first step toward greater business influence on education? A world in which children from low-income neighborhoods paint murals that glorify brands?

Gerald Levin, former CEO of AOL-Time Warner, claimed at the time of the recent announcement of the companies' planned merger that "this is not just about big business. This is not just about money. . . . This is about making a better world for people."[15] Yet how can the AOL merger possibly advance the good of the world? Especially given that it couldn't even protect its own stock price, which has fallen by 80 percent since the time of the merger. The reality remains that the business community will never place good customer service, ethical trading, and social investment above moneymaking whenever the two come into conflict. Both the political choices and all the other actions of businesses will be made in the primary interests of the shareholder value and profit projections, not justice, equity, or morality.[16] In fact, given the current state of the law as regards duties to shareholders, Anglo or American businesses *have* to do so. In the United States investors can sue a firm if they believe that it is not making the best use of their money by putting it into social causes. When Kodak gave $25 million to a black civil rights organization in Rochester, New York, it was threatened with a shareholder suit and had to withdraw its donation. Corporations can only do good if they can show that they can make more money by doing so.

So the oil companies jumped onto the environmental bandwagon, not for any ethical reasons—although the principals involved may also have believed that this was the right thing to do—but for clear commercial gain. BP viewed climate change as a "source of business risk" and felt that by changing company policy it would gain a distinctive "competitive advantage" and win favor with customers, regulators, and legislators, and thus bring about future business benefits.[17] Diamond cartel De Beers would have been unlikely to have instructed its suppliers to stop purchasing "blood diamonds" from Sierra Leone as it proposed in July 2000 if the UN Security Council had not banned trading in such gems until a certification process could be set up, or if it had not feared a consumer boycott of diamonds similar to that which harmed the fur industry in the 1970s and 1980s.

For this is not about ethics, this is about business. Sometimes both sets of considerations will coincide, but clearly not always. Corporations are not society's custodians, they are commercial entities that act in the pursuit of profit. They are morally ambivalent: Siemens, Knorr, Deutsche Bank, AEG, BMW, Volkswagen, and Daimler-Benz have all been cited by the Holocaust Education Trust as exploiters of slave labor from the Nazi concentration camps. Governments on the other hand are supposed to be social institutions, within which responsiveness to citizens is, or at least should be, central. Downgrading the role of the state in favor of corporate activism threatens to make societal improvements irreversibly dependent on the creation of profit. Standing back while corporations take over, without being willing to set terms of engagement or retain the upper hand, leaves governments increasingly in danger of losing the support of the people still further. Remaining mute in the face of the Silent Takeover risks leaving us ultimately without either recourse or representation.

11

Reclaim the State

The Rise of Protest

As we have seen, while the power and independence of governments wither and corporations take over ever more control, a new political movement is beginning to emerge. Rooted in protest, its advocates are not bounded by national geography, a shared culture or history, and its members comprise a veritable ragtag of by now millions, made up of NGOs, grassroots movements, campaigning corporations, and individuals. Their concerns, while disparate, share a common assumption: that the people's interests have been taken over by other interests viewed as more fundamental than their own—that the public interest has lost out to a corporate one.

The protesters include ordinary people with ordinary lives: housewives, schoolteachers, pensioners, students, businesspeople, suburbanites, and city folk: blue- and white-collar workers alike. Although their goals may be divergent and may even at times be at odds with one another, they share a skepticism about the promises and assurances given by those in authority and—thanks to the neoliberal orthodoxy which has taught them that the state cannot solve their problems—considerable uncertainty about the role of government.

The apparent inability or unwillingness of our elected representatives to defend our interests against those of business has created a cycle of cynicism. People do not look to government to solve their problems, and politicians therefore have little to lose if they focus their attention on business rather than on voters. Low voter turnout, falling levels of trust, and increasingly visible corruption have contributed to a widespread feeling that politics simply does not matter. It is almost as though both sides of the electoral equation have given up on democracy, through a suspicion that elections don't really change anything substantial. In a world where governments are proving less effectual than corporations, trust in representative government is at an all-time low. Traditional deference to politicians, along with many other experts, has evaporated, leaving a citizenry that increasingly demands an effective and decisive say in important issues, a say that seems ill served by the electoral ballot box.[1]

These protesters believe that taking the chance that what is good for business is good for us and our communities is just too high a risk, hazarding the food we eat, the environment, and the democratic process. While some may welcome the recent attempts of various corporations to address some of the failings of the system and contribute to the social sphere, they tend to see these attempts as window dressing or corporate PR, and remain skeptical about companies' motives. At the same time they reject representative government as an ineffective, coopted, and flawed mechanism for dealing with the failings of the market or representing their interests on the global stage, and reject the politics of today as the "politics of Narcissus," concerned only with presentation and "spin." They choose to voice their concerns on the street, on the Internet, and in the shopping malls, because they feel that these are the only places that they can be heard. They will not trust either government or business except in terms of responsiveness and results.

Pulitzer Prize–winning author Thomas Friedman's "McArches World,"[2] in which countries with McDonald's in them do not go to war, is being replaced by a McConflict world, in which wrecked McDonald's shop fronts have become a symbol of the discord within and José Bové, the French farmer who destroyed a McDonald's construction site, has become a folk hero. International order provided courtesy of multinational corporations may have been purchased at the price of domestic anarchy.

But it is not just brands that the protesters attack—governments and multilaterals are targeted with equal ferocity, often with positive effect. There was the forcing through of a pact on global climate at the Earth Summit in Rio de Janeiro in 1992, where NGOs "set the original goal of negotiating an agreement to control greenhouse gases long before the governments were ready to do so, proposed most of its structure and content, and lobbied and mobilized public pressure to force through a pact that virtually no one else thought possible when the talks began."[3] And the collapse of the Multilateral Agreement on Investment (MAI), thanks to the efforts of numerous consumer groups and environmentalists who feared that the draft treaty to harmonize rules on foreign investment would have disabled national governments' ability to protect their own citizens in the face of corporate demands. Then there was the balaclava-wearing leader of the Zapatistas, pipe-smoking subcommander Marcos, who in 1994 waged cyberwar against the pro-NAFTA Structural Adjustment Policies of Mexico's PRI government. And the Pink Fairies of Seattle, Prague, Gothenburg, and Genoa who in their Bo-peep outfits, pink bras, Lycra, sequins, and wings helped to disrupt recent IMF, WTO, EU, and G8 meetings. Jubilee 2000 successfully pushed for a dramatic reduction in the debts of the poorest countries. And the presidential building in Quito, Ecuador, was taken over in protest in January 2000 against President Jamil Mahaud's austerity programs. A culture of protest is emerging that threatens to overturn the status quo.

Through demonstrations, publicity campaigns, and direct action schemes, the protest movement attempts to raise the costs to businesses and governments of continuing with whatever practices protesters consider damaging and to shape the terms on which the new elites can operate. As journalists, academics, activists, and ordinary citizens speak out against the omnipotence of big business and the unreliability of government, nowadays protesters log on rather than turn on and protest rather than drop out.

What makes this movement particularly remarkable is the breadth of its appeal, and the extent to which it has managed to coalesce divergent interests. In February 2003 alone, for example, we saw Greenpeace activists abseiling onto Exxon U.K.'s company headquarters, a hundred Exxon gas stations in Britain closed as activists dressed as tigers chained themselves to pumps, anti-fur campaigners back invading the catwalks, Minnesotan pensioners boycotting GlaxoSmithKline for its attempts to stop its drugs from being sold in the U.S. at the cheaper prices it sells them in Canada, and millions of SUV windshields slapped with faux parking tickets saying "Violation: Earth" by American anti-SUV activists. Traditional and nontraditional groups are working together in unprecedented ways to achieve solutions, rather than, as in the past, seeing each other as part of the problem.

The scandal of BSE (mad cow disease) in Britain, for example, was significant in the extent to which it gave former enemies a common cause: "Civic association—the classic expression of civil society—and uncivil politics—the presumed expression of anomic democracy—joined hands against government untrustworthiness. Farmers and producers, environmentalists and consumer groups, opposition politicians and newspapers mixed conventional forms of participation with social activism in response to untrustworthy government."[4]

The debate over genetically modified foods elicited a similar response, except that agrochemical corporations joined politicians as the focus for protesters. In Britain guerrilla gardeners—environmental activists whose tactics included night attacks on GM crops—found themselves sharing a platform with the Women's Institute, a traditional bastion of British conservatism, in condemning GM foods.

At the World Trade Organization talks in Seattle in November 1999, a similar range of divergent interests gathered outside the convention hall to express their concerns over international free trade. Trade unionists, environmentalists, and anarchist groups differed in their goals but shared a common hostility to the way that global markets were being sliced up and controlled by the most powerful governments and corporations. The image of these erstwhile enemies holding hands symbolizes the extent to which civil society is now speaking with a common voice, at least so far as it shares common concerns. Protest is becoming institutionalized as an acceptable form of expression.

The movement has no fixed membership, so it can mobilize support around shared concerns, national or global, as and where appropriate. This lack of permanent mass membership and of a physical base does not weaken it, rather it makes the movement more flexible and able to tackle diverse issues, many of which may cross national boundaries. Its power is widely distributed: "One does not need an army, control over governmental bureaucracies, massive wealth, or even large numbers of activists to be effective."[5] In the age of the Internet mass action can be orchestrated with unprecedented ease. Sharing information and strategies and building links is easier and cheaper than ever before. We saw the pressure corporations are now under from e-boycotts. Similarly "A draft of the MAI text, posted on the Internet . . . allowed hundreds of hostile watchdog groups to mobilize against it. [And] the Seattle trade summit was disrupted by dozens of websites which alerted everyone (except, it seems, the Seattle police), to the protests that were planned."[6]

As the power and credibility of politicians wanes and the power of corporations and international organizations grows, the protest movement has been gaining momentum. A hundred NGOs turned up at the WTO Ministerial Meeting in 1996; three years later, in Seattle, there were over a thousand.[7] Over 100,000 Bolivians took to the streets in February 2000 in protest against their government's decision to privatize the national water supply. The Washington-World Bank/IMF protests that April were attended by over ten thousand demonstrators. At the end of June 2000 forty thousand people gathered in France outside the court at which José Bové was being tried. In July a consumer boycott by thousands of Japanese housewives brought down Sogo, the Osaka retailer that had come to epitomize business-government cronyism in Japan.[8] Twenty thousand protestors converged on Prague in September; and at the EU summit in Nice in December that year 100,000 turned out.

In 2001 tens of thousands rose up against IMF plans in Ecuador in February; eighty thousand took to the streets in April in Quebec against the Free Trade Area of the Americas agreement; thirty thousand protested in Gothenburg against the IMF summit; more than 150,000 in Genoa in July; and then in December more than 1 million Argentineans poured onto the streets of Buenos Aires in protest against the economic austerity measures that were pushing two thousand Argentineans below the poverty line each day.

And then in January 2002 in Colombia, a thirty-five-day occupation of the offices of the state water, telecoms, and electricity company by workers after the Colombian government announced plans to sell the enterprise; in March thousands of South Africans on the streets demanding that the state power company stop cutting off electricity to those too poor to pay; in April 2002 in India, a nationwide strike of over 10 million workers over labor reforms and the government's privatization plans; in Uruguay in May, a general strike protesting against the government's economic policies and a bill which aimed to raise taxes on salaries and pensions; in November, 300,000 demonstrators gathering in Florence for the European Social Forum; and in Argentina in December, thousands of protestors marching through Buenos Aires, to mark the anniversary of the 2001 riots, hurling paint bombs at the stock exchange and protesting against IMF policies.

These are early days, but if people continue to feel alienated from traditional politics and distrustful of the politicians' agendas, if they continue to feel abandoned by the state, and increasingly of the opinion that politics has been coopted by business, if people continue to feel that the only real power is in the hands of unelected institutions—huge corporations and unaccountable supranational structures—the voice of protest will only grow louder and we will continue to see a shift from the politics of acceptance toward that of dissent. In the nineteenth century workers, and in the early twentieth century women, protested to get the vote; today protest centers on the assumption that their votes have become insignificant.

Protest as the Catalyst for Change

As we have seen, in today's world economics has become the new politics, and the pursuit of economic goals now outweighs political and social concerns. Governments pursue market share rather than territorial gain, and politicians depend on big business to fund their campaigns and provide the jobs they need to win elections, threatening what impartiality they once had. At the same time people have become increasingly distanced from politicians, and politicians have shown themselves equally out of touch with their electorates. Even James Wolfensohn, head of the World Bank, now concedes that "globalization is not working at the level of the people,"[9] and it is clear that wealth is not trickling down as has been predicted. Meanwhile, the IMF's 2,700 employees dictate economic terms to 1.46 billion people, and corporations are now openly in the business of politics. The democratic deficit is fast becoming a democratic chasm, and protest is emerging as the only way for other voices to be heard.

The protest movement gives a voice to people who have been denied the right to elect their governments, as well as to people who no longer feel that their representatives are acting on their behalf. It empowers people who otherwise would have no recourse, in particular the young, the group throughout the world's democracies least likely to express themselves through the traditional ballot box. By rejecting traditional notions of representative democracy, it makes democracy more direct and puts it in people's own hands. By questioning, criticizing, and publicizing, it "can change the terms of disclosure, and the balance of different components in the international constellation of discourses."[10]

The movement's success has given participants a sense of empowerment, and demonstrated that there are alternatives to the frustration and alienation that many experience. They have proven that the *demos* has a clear role to play in this commerce-centered world, in applying pressure to society's decision makers, in making democracy more robust, if more uncertain.

In a world in which ideology competes with ice cream and the policies of the dominant parties are almost indistinguishable, so that there is no apparent gain from changing the government, protest places on the agenda policies that the dominant parties would not otherwise offer the electorate. In a post–cold war era in which the United States has become the only "imperial power," we see a rise in popular dissent, because people see no alternative but to take issues into their own hands.

Of course, such protest does not provide a long-term solution to the Silent Takeover. Its limitations mirror those of consumer activism—unsurprisingly so, given their shared genesis in the discontent of the early 1990s and their similar methods of expressing discontent. The commonality of interests often centers on a shared general disillusionment, rather than specific concerns or proffered solutions. In some cases protesters are motivated by a sense of common good; but in others they are concerned only with safeguarding their own interests, or those of a limited group—the "raise less corn and more hell" variety of protest, like the British fuel protests of autumn 2000. As we have seen, pressure groups need to play to the media, which encourages polarized posturing, the demonization of "enemies," the oversimplification of issues, and the choosing of fashionable rather than difficult causes to champion. Issues such as soil erosion, nitrate leaching, and forest biodiversity in Africa hardly ever get a mention. And the need for media attention can inspire violence. As Brian, the American student I met en route to Genoa, put it, "There has to be trouble, otherwise the papers won't report it, we won't get our concerns on the front page."

Various pressure groups that play a large role in civil society have taken up the mantle of people's champion, yet they lack any sort of democratic mandate, are often narrowly focused on the priorities of their members, or of their leadership, and may work to impose their values irrespective of those of others. Some aim to speak for the poor and the marginalized, but not all. Because they concentrate on single issues, they may feel no need to concern themselves with the problems of others, as would occur in a genuine democracy. Sometimes the coalitions of interests are global in their concern, but often they have highly nationalistic undertones. And sometimes the wishes of the *demos* can be downright nasty, like the British hysteria about pedophiles, largely stirred up by a corporation, News International, through the pages of its *News of the World* tabloid newspaper and resulting in such fiascos as that of the Bristol pediatrician who had to go into hiding because the mob couldn't tell the difference.

Protest is far removed from any familiar notion of participatory democracy. For those who are not prepared to stand in the clouds of tear gas outside another intergovernmental conference, or to live in tunnels beneath proposed road development sites, the scope for involvement is limited. We can post off a check once a year to Greenpeace—or, like the majority, sit back and watch the dramas unfolding on our television screens, unsure whether we really identify with or support the tactics of the protesters. Can these masked masqueraders really be representing our views?

Protest acts as a countervailing force to the Silent Takeover, yet because it is not fully inclusive it shares, to a degree, the illegitimacy of its opponent. The institutionalization of protest risks leaving us with a political system where those with the most intensely held opinions, those who shout the loudest or are the best organized, are the people to whom politicians and CEOs respond. Antiabortion campaigners in the United States and defenders of foxhunting in the U.K. have distributed preprinted postcards to group members for them to sign and send to local representatives. E-mail allows pressure groups to mobilize thousands of members instantaneously, who can shoot off standard forms of protest to express concern about a single issue. The silent majority risks becoming disempowered by the vocal minority. Corporations risk being tried by kangaroo courts, while politics risks being permanently assigned to an arena in which the battle for political sway is fought on the one hand by corporations and on the other by pressure groups, with ordinary people's interests lost in the struggle.

But despite the limitations of protest, despite its failure to balance effective means with democratic ends, despite the fact that it can never by itself be a long-term solution, the question remains as to whether, as its power increases, it will be able to act as a catalyst for reform. Can protest change politics in the same way as it is beginning to change the corporate agenda? Can protest pressure governments into once again putting the people's interests first? Can it force politicians to return to true democracy and provide the stimulus for them to come up with genuine interparty debate and politics that will mobilize voters? Can protest act to reestablish government as a democratic forum within which different social needs are weighed, and all is not reduced to either the corporation or the individual? Can protest serve to reinvent the state?

Power to the People

History suggests that it can. Democratic governments and large corporations alike need mass support to survive, and this gives an enormous power to the people to impose their own terms of cooperation.

At the beginning of the last century the United States was enjoying a period of relative prosperity not dissimilar to our own: Over the previous thirty years agricultural production had more than doubled; the production of coal had gone up five times; and that of crude petroleum had increased twelve times.[11] But farmers were facing increasingly low returns despite increases in output, industrial production was rapidly becoming concentrated in the hands of a few large corporations, and powerful political bosses at city, state, and national levels were giving favors to big business to win financial support for their machines. "High tariffs, business monopolies, inequitable taxation, and civic graft"[12] ate into people's incomes.

News magazines revealed cases of serious and widespread collusion and corruption involving business and politics.[13] Articles exposed the threat to public safety from contaminated food, and the extortion practiced by the railroads. Newspapers overflowed with editorials deriding John D. Rockefeller, Andrew Carnegie, Jay Gould, and other "robber barons."

An increasing proportion of the American people began to feel vulnerable, uneasy, and angry.[14] There was a grassroots awakening, as the realization that corruption, monopoly franchises, and discriminatory pricing were adversely affecting the public interest spread.[15,16] Urban and rural interests combined in a rare alliance, and a public mood emerged "generally sympathetic to calls for reform."[17]

There was an increase in political activism, but it was outside traditional political channels.

> Prior to 1900, it was rare for people to turn away from their political parties and find methods other than the vote for influencing government. Beginning in the early twentieth century, this older structure of political participation gave way to new patterns. Voter turnout fell, ticket-splitting rose, and relatively few voters could be counted on to support the regular party candidates year after year. In the same period, a great variety of interest groups successfully pioneered new ways of influencing the government and its agencies.[18]

Citizens' organizations were formed to improve urban conditions, tackle the squalid living conditions of the poor, and address abuses of political and corporate power.[19] Groups were formed to demand rights for women or for workers, and consumers increasingly refused to buy products made by child labor. Voters demanded new powers, and popular pressure led to the introduction of various legislative reforms: the Corrupt Practices Act, which addressed the illicit relation between money and politics; the direct primary, which put the choice of political candidates in the hands of voters rather than those of the party machines; the initiative, which allowed citizens' organizations to propose legislation; the referendum, allowing citizens a vote on crucial state laws; and the recall, which allowed the removal of corrupt or incompetent officials before the end of their term of office.

Politicians realized that they could no longer depend on people to vote unthinkingly along traditional party lines, and had little choice but to respond. A third party entered the 1912 electoral race: the Progressive Party, with Theodore Roosevelt as its candidate, a man whose robustly anti-corporate pronouncements—"to destroy this invincible government, to befoul the unholy alliance between corrupt business and corrupt politics, is the first task of the statesmanship of the day"—would be most welcome today. This party's success—polling 25 percent of the vote in a system which remains notoriously unwelcoming to third parties—led all parties to rally against the corporate domination of politics. Business was systematically regulated for the first time, and new forms of political participation emerged that downgraded party bosses and gave more powers to ordinary voters.[20]

This combination of protest with voter and consumer pressure, and the skillful use of the media, gave an irresistible momentum to the Progressives' cause. Progressivism became "the first (perhaps the only) reform movement to be experienced by the whole American nation. Wars and depression had previously engaged the whole nation's attention, but never reform."[21]

Fifty-odd years later, in the wake of the Vietnam War, we saw a rebirth of radicalism, but this time internationally. Demands for peace, the extension of civil rights, resistance to racism, and the emancipation of women were just some of the issues that produced unparalleled scenes of protest in London and Washington, and took France to the edge of revolution in May 1968.

And it was in this context that in Europe a similar process to that which took place within the Progressive Era was played out, a process that translated broad societal disaffection into swift and effective action outside traditional political channels, focused this time not on social and political issues but on the environment.

In the 1960s declining levels of economic growth in combination with the new environmental problems that were facing advanced industrial democracies—nuclear power, resource shortages, toxic waste, acid rain—led to a growing sense of public concern about the environment. Scientific evidence of these concerns became available for the first time, making grimly clear the environmental and health consequences of industrial development. Front-page headlines catalogued disasters: oil spills that damaged aquatic life and the coastlines of California and Cornwall; and toxic waste that spilled into the Rhine, killing fish and poisoning drinking water. Rachel Carson's *Silent Spring*[22] galvanized popular support with a bleak depiction of a miasmatic world in which, thanks to the overuse of insecticides and pesticides, birds no longer sing.

An environmental movement emerged to fight against what was seen as corporate and political negligence. Like the Progressive movement before it, it eschewed conventional, hierarchical organization by party or interest group, and it rejected traditional forms of participation and the idea that the vote alone was an adequate form of political expression. Again like the American Progressives, it built its success on successfully harnessing latent public anxieties and mobilizing the broad coalition of interests that gathered together under the environmental banner. In their shared goal of protecting the planet, mainstream conservationists of the "save the whale" type united with radical groups who opposed industrialization and preached apocalyptic messages about global destruction. Disasters such as the Chernobyl nuclear explosion in 1986 and the discovery of the link between skin cancer and the hole in the ozone layer were used to maximum effect.

Consumers stopped buying products containing CFCs and boycotted products tested on animals. Citizens' groups were formed to address local environmental problems. People joined organizations such as Greenpeace to protest nuclear testing, ozone depletion, and global warming. Membership of ecology groups in Britain rose by the mid-1980s to half a million members, and groups in both West Germany and the Netherlands had membership levels of around a quarter of a million.[23] Direct action and protest were seen as the only ways to promote the environmental cause.

And the strategy worked. Despite the success of New Right governments that came into power in much of the developed world in the early 1980s, for whom the environment was not a natural cause to champion, the environmentalists doggedly managed not only to keep the issues before the public but also, in a relatively short time, succeeded in transforming the political landscape. Soon the groundswell support for environmental issues led existing political parties to compete in demonstrating their "green" credentials, and European leaders of both left and right—Mitterrand, Kohl, Thatcher—began to claim to be environmentalists.

New political parties were established which explicitly advocated a green agenda. "By the beginning of the 1990s, green parties or their New Left supporters had won seats in the national Parliament or EC Parliament for most nations in Western Europe."[24] These parties showed a remarkable ability to break the left-right mold of the established party systems and, as in the case of the American progressives, appealed to voters who had previously followed a strict partisan line. They still remain significant players in continental European politics. Green parties are currently members of governing coalitions in Germany, France, Belgium, Italy, and Finland. In Britain they maintain a strong presence in local government.[25] The greens are the fourth-largest coalition in the European Parliament. Green issues are now firmly established on the political agenda.

Whither the Takeover?

The combination of consumer activism and political protest can clearly be extremely effective. And the parallels between the eras of the

American Progressive movement, the European environmental movement, and today are striking: a comparable disillusionment with government and a rejection of mainstream politics, a suspicion of big business, a willingness to take to the streets and protest, the use of consumer pressure as a political and economic weapon, a broad coalition of interests; avoidance of traditional economic debates in favor of debates over quality of life issues, the ability to combine idealism with self-interest, and a nonpartisan- and nonclass-based orientation.

But while the successes of the Progressives and the environmentalists teach us that governments *can* respond, the question of whether they *will* respond to protest against the Silent Takeover remains unanswered. The Internet has revolutionized the speed at which businesses respond to activism—an internal memo to senior executives of a large multinational in April 2000 pointed out that "riots in Seattle, Washington, D.C., and London on May Day are all evidence of rising tensions which we ignore at our peril"—yet national institutions lag behind, slowed down by traditional structures that can seem ill suited to the new millennium. Will this groundswell, these acts of vigilantism and anger, kick start a reform of the twenty-first century's political agenda in the same way that these earlier movements did?

To a limited extent they already are. As the rhetoric of business has changed, so has political discourse, at least among parties of the center-left. Clinton's sympathetic comments on the Seattle protesters,[26] the attempts in the 2000 election by Gore to woo the green vote, Labour's "End Rip-off Britain" campaign, Lionel Jospin's calls for a "World Environment Organization" to counter the thrust of the World Trade Organization. And then, after September 11, Clinton's speech in Belgium in October 2001 in which he talked about the "unacceptability of a world in which there is one set of rules for the rich and another for the poor"; the call by Guy Verhofstadt, Belgian prime minister and president of the European Union, for "global binding agreements on ethics and the environment," New Labour's announcement that they would double the amount of British aid to least-developed countries. Changes in rhetoric that suggest perhaps a fear of rising popular dissent and a dawning on politicians that the pendulum of global capitalism may have swung too far, that the rising tide never did manage to lift all the boats, that corporations may be too powerful, that inequality may be responsible for an unacceptably high degree of social unrest, and that it is time to lay the *capitalismo selvagem* (savage capitalism) of Reagan and Thatcher to rest. Social justice may be coming back into political fashion—among left-of-center parties, at least.

But even here, the extent to which these parties are capable of translating this new rhetoric into a sustainable reality remains in question. Will others in the Democratic Party take up Clinton's radical new words? Politicians in office rather than those already out? Guy Verhofstadt's call for global binding agreements has so far been ignored. And Britain's proposed post–September 11 "Marshall Plan for the World" was summarily dismissed by the U.S. government.

For in the United States under Bush there has been no attempt even to address any of the concerns raised by the protest movement. Instead, under George W we have witnessed the promulgation of a brand of conservatism that has proven far from "compassionate" and the escalation of policies that put the interests of corporate America first: the $350 billion in tax cuts which will benefit the rich at the expense of the poor; the scaling back of environmental regulation and the proposed opening up of the Arctic National Wildlife Refuge for exploration; the deregulation of business, especially the oil industry; Tom DeLay, Republican majority whip of the House, telling airline lobbyists that they had to back the Republican position of not federalizing airport security on "ideological" grounds; an economic stimulus plan that focused on tax cuts for big business rather than on increases in public spending; and the War on Iraq cynically used by the administration to avert media attention from its inaction on addressing corporate corruption. As of April 2003 no convictions had been handed out to any of the leading white-collar criminals.

Not only is Bush threatening to remove all the great progressive gains of the last three decades, he is also threatening to undermine cooperation in the world with his complete unwillingness to embrace multilateralism—absolutely essential to secure human rights, national security, and the sustainability of our environment, and to keep global capitalism under control. The War on Iraq waged without the support of the United Nations, Bush's downgrading or junking of humanitarian interventions, his refusal to ratify the Kyoto Protocol on climate change, his unwillingness to sign up to a draft agreement updating the 1972 Biological Weapons Convention, his opposition to the International Criminal court, and his refusal to ratify the Small Arms Treaty because of the interests of American-based arms manufacturers, to name but a few. The country in which the dangers posed by the Silent Takeover are most apparent is the country whose reigning government is promoting the takeover itself.

That is the short-term perspective on change. But if governments are not farsighted enough to confront the Silent Takeover; if they are not prepared to learn from the lessons of the Progressive and environmental eras, to search for solutions, to resist pressure from big business when the market mechanism fails or when the pursuit of corporate profit is against the public interest, and use their coercive powers to demand compliance from corporations; if they remain out of touch with the public and do not work to give people a greater say in the system, using new technologies to consult voters and allow increased levels of participation; if they forget that people will no longer support a world which is exclusively about growth rates and private capital flows and in which inequalities continue to deepen; if they fail in all this, they will sign their own death warrant, and the world we live in will be one in which corporations rule, markets are above the law, and voting becomes a thing of the past. The final stage of the Silent Takeover is the end of politics itself, collapsing into cycles of protest, repression, and despair.

The New Agenda

But can politics be reframed so as to avert this nihilistic scenario? Is there a new agenda that could be embraced that could rebuild democracy for the people? Can social injustice, inequality, and power asymmetries be addressed so as to make politics a product once again worth buying, and can globalization be made to work for all and not just the few?

I believe that they can, that a new agenda is possible, based on principles of inclusiveness, a reconnection of the social and the economic, and a determination to ensure that everyone has access to justice wherever they are. And that what has been preventing its birth has not only been the safeguarding of special interests or a lack of resources: It has been a lack of moral imperative, responsibility, or political will.

First, at the national level, this new agenda necessitates a disenfranchisement of corporations. Corporate funding of political parties and election campaigns makes a mockery of democratic principles, and continues to ensure that politics remains skewed toward the interests of the few—exclusive rather than inclusive. In practical terms this means breaking the financial stranglehold corporations have on politics, and a commitment by those governments that have not already done so to introduce reform of political financing and public funding of election campaigns. Any private funding of election campaigns will always come with strings attached. If trust is to be restored, politicians will have to prove to the electorate that they are working for the public, and not a private, good.

Second, the steadfast belief in trickle-down economics, the legacy of Reaganomics and Thatcherism, an axiom which has been used to justify everything from corporate welfare to tax cuts for the rich, must once and for all be laid to rest. Growing inequalities and corporations' tendencies to capture the gains from subsidies or tax cuts for themselves with no regard for the local community provide glaring disproof of the trickle-down theory. In practical terms the rejection of this axiom will necessitate not only the scrapping of the policy of corporate welfare but also a rethink of redistributive tax policies and public expenditure. A world of gated communities next to ghettos is not only unconscionable, it is also dangerous. The "free lunch" school of politics, in which politicians make inflated claims and generate inflated expectations without admitting that tradeoffs will undoubtedly be needed, must be laid to rest.

And third, the power of corporations at a national level must be checked. Reregulation rather than deregulation, and even in some cases the contemplation of the federalization of public goods. Stronger antitrust bodies, with the increases in funding that will be needed to support them. Cross-ownership restrictions on media enforced. Mandatory reporting requirements on issues relating to the environment and society. And the integrity of information and research ensured: obligatory disclosure of potential conflicts of interest, and corporate sponsorship of the public realm made subject to stringent controls. Without a strong regulatory framework in place, the market becomes a free-for-all, too often at our own and our neighbors' expense.

But reframing politics at the national level, though necessary, is not sufficient. In a world of global capital politics must be reframed at the global level, too. This will entail addressing the dominance of trade and corporate interests in the global sphere, as well as the question of how to best meet the needs of those who have not benefited from globalization.

To this end we will need first to put in place mechanisms to help people fight against injustice as part of a wider political rebuilding of institutions. All people, wherever they are, must be extended the rights we in the North take for granted. Workers and communities everywhere must be guaranteed basic rights to minimum health, safety, and welfare standards at work and not be dismissed or dispossessed without adequate compensation. Multinational corporations must not be allowed to infringe these rights, wherever it is that they operate.

A world in which people have no access to justice is one in which discontent will continue to fester. So it is imperative that we ensure that the perpetrators of corporate injustices be held to account, wherever they are, and that their victims have redress whoever they are. In the long term this is a matter of strengthening both local and international regulation of companies and making enforcement effective. In the short term there are two clear initiatives that can be taken.

First, governments of the North must commit themselves to legislative reforms that will ensure that the corporate veil can be pierced and parent companies be held responsible for the actions of their subsidiaries in whatever country they operate. Second, workers and communities everywhere must be given access to a global legal aid fund.

Next, we need to set up a World Social Organization (WSO): an organization which will counter the dominance of the World Trade Organization and will establish rules and regulations that will reframe global market mechanisms to ensure the long-term protection of human rights, labor standards, and the environment. Such an organization must have teeth as sharp as those of the WTO and equally effective powers of enforcement. Together with the WTO, it will be subject to a new adjudication mechanism that will seek to reconcile trade and other interests when the WTO and WSO clash, as they undoubtedly will, so as to best serve the public good.

But we in the North must be careful not to use this new organization as a form of protectionism. The developed world must help developing countries meet the costs of better global standards, and the different starting points of different nations must be taken into account when designing new protocols.

And finally, we must address the problem of alleviating the positions of those who are most excluded and marginalized, the losers from globalization. At least, we must cancel debt and reverse the outflows of capital from the South to the North. We must significantly increase overseas aid, which for the least developed countries has fallen 45 percent in real terms since 1990, and we must rethink the ways in which it is delivered. And we must pull down all trade barriers on agricultural and textile products from the developing world—developing countries are losing almost $2 billion a day because of inequitable trade rules. The commitment at the Doha round of the WTO to enter negotiations on the issue of agricultural subsidies is frankly not good enough.

But more than this, we will also need new money to realize our new goals. The world needs a new global tax authority, linked perhaps to the UN system. The authority should have the power to levy indirect taxes, for example, on pollution and on energy consumption, which can then be spent on protecting the environment. The authority will also need to be able to levy direct taxes on multinational corporations, in order to fund the development of global environmental, labor, and human rights norms. And specific health taxes on tobacco and alcohol companies should be levied to fund a global health fund.

These six steps are only the beginning of an agenda for action to recast globalization. They are not the only steps we could take—of course not. But they are a way to begin to reunite the global economy with social justice, a way to begin to address the fundamental concerns highlighted by the Silent Takeover.

A better world is possible: a world of greater equity, justice, and true democracy. But here is a warning: Unless those in power do address these issues, those disenfranchised by the Silent Takeover or those who chose to speak for the disenfranchised will keep on trying to batter down the doors of power, in whatever ways they see most fit. If we continue as a world to perpetuate such power asymmetries, and if inequalities continue to grow at the rate we have seen them grow over the past twenty years, what we will see is a replacement of politics by protest, an institutionalization of protest and rage, and with it the demise of democracy itself, even in those nations that pride themselves on being democratic. Until the state reclaims the people, the people will not reclaim the state. Until the benefits of globalization are shared more widely, people will continue to rise up against globalization.

NOTES

Chapter 1

1. Of particular note is the fact that the World Bank's *World Development Report of 2000* acknowledged that inequality was bad for growth, economic growth does not automatically reduce poverty, and that poverty is not simply an economic problem but also a political one.
2. Interestingly, the death row campaign badly misfired. After victims' rights groups picketed a Texas Sears store in February 2000, Sears, the second-largest retailer in the United States, dropped its contract with Benetton for a private label line. The line had been expected to generate some $100 million in sales in its first year. Olivero Toscani, Benetton's longtime creative director, was another casualty of the campaign. A joint statement with Benetton owner Luciano Benetton implied he left to "take on new projects."
3. According to the World Bank's 1999 *World Development Indicators,* there were 654 television sets per 1,000 people in highly developed countries. In the USA this is as high as 847 per 1,000.
4. Ben Bagdikian, *The Media Monopoly,* 4th ed. (Boston: Beacon Press, 1992), p. 157; "Buy-Nothings Discover a Cure for Affluenza," *The New York Times,* November 29, 1997.

5. E. Herman and R. McChesney, *The Global Media: The New Visionaries of Corporate Capitalism* (London: Cassell, 1997).
6. www.adbusters.org; see also:www.oneworld.org/ni/issue278/jamming.html.
7. "Shoppers Are Pigs," *Wall Street Journal,* November 19, 1997. See also: "Consumer Republic," *Adweek,* November 22, 1999.
8. Anderson and Cavanagh, "Top 200: The Rise of Global Corporate Power," Institute for Policy Studies (Washington, D.C., 1999). This is calculated by comparing GDP to company sales. Even if GDP is compared with value added, the economic might of corporations remains striking. For instance, General Motors, using the value-added measurement, emerges as the fifty-fifth largest economy in the world! (UNCTAD 2002).
9. Compiled from country and company data from *The Economist* and *Fortune 500* lists, 2000.
10. "Business: Rupert Laid Bare," *The Economist,* March 20, 1999.
11. Stephen Byers speech transcript, Mansion House, February 2, 1999; http://www.dti.gov.uk/ministers/archived/byers02021999.html.
12. *Pulling Apart: A State-by-State Analysis of Income Trends,* Center on Budget and Policy Priorities and the Economic Policy Institute (2000).
13. Robert Reich, *The Future of Success* (London: Heinemann Press, 2000), p. 97.
14. Ibid., p. 98.
15. Federal Election Commission Data, December 18, 2000.
16. Robert D. Putnam, Susan J. Pharr, and Russell J. Dalton, "Introduction: What's Troubling the Trilateral Democracies," in Putnam and Pharr, eds., *Disaffected Democracies* (Princeton: Princeton University Press, 2000), p. 14.
17. Robert Puttman, *Bowling Alone* (London: Simon & Schuster, 2000).
18. "Too Much Corporate Power," *Business Week,* September 11, 2000.
19. Remarks by Eric Hobsbawm in "Roundtable: The Global Order in the 21st Century," *Prospect* 44 (August/September 1999).
20. J. Stiglitz, "The Economic Consequences of Income Inequality," Federal Reserve Bank of Kansas City (1999).

Chapter 2

1. William Greider, *One World, Ready or Not: The Magic Logic of Global Capitalism* (New York: Simon & Schuster, 1997), p. 362.
2. Chris Pierson, "Social Policy," in David Marquand and Anthony Seldon, *The Ideas That Shaped Post-War Britain* (London: HarperCollins, 1996), p. 151.
3. Desmond S. King, *The New Right: Politics, Markets and Citizenship* (Homewood, IL.: Dorsey, 1987), p. 58.
4. Nigel Lawson, "The Frontiers of Privatisation," speech to Adam Smith Institute conference on privatization (1988), quoted in Peter Riddell, *The Thatcher Era and Its Legacy* (Oxford: Blackwell, 1991), p. 6.
5. Desmond S. King, *The New Right,* p. 68.
6. David Marquand, "Moralists and Hedonists," in Marquand and Seldon, *The Ideas that Shaped Post-War Britain,* p. 14.
7. Ibid., p. 15.
8. "The Thatcher Record," *The Economist,* November 24, 1990; Economic Report of the President, January 1993, Table 13–59, p. 462; "The Trouble with Theories: Assessing Reaganomics," *The Economist,* January 21, 1989.
9. John Gray, *Beyond the New Right* (London: Routledge, 1993), p. vii.
10. John Moore, social services secretary, speech in London, September 26, 1987, quoted in Peter Riddell, *The Thatcher Era and Its Legacy,* p. 127.
11. Peter Riddell, *The Thatcher Era and Its Legacy,* p. 150.
12. Ibid, p. 151.
13. Quoted in Peter Jenkins, *Mrs. Thatcher's Revolution* (Cambridge: Cambridge University Press, 1987), p. 326.
14. Ibid.
15. Richard P. Nathan, "The Reagan Presidency in Domestic Affairs," in *The Reagan Presidency: An Early Assessment,* Fred. I. Greenstein, ed. (Baltimore: Johns Hopkins University Press, 1983), p. 49.
16. Philip Morgan, "The Privatization of the Welfare State: A Case of Back to the Future?" in Phillip Morgan, ed., *Privatization and the Welfare State* (Aldershot, UK: Dartmouth Publishers, 1995), p. 12.
17. Ibid., p. 152.
18. See for example, Peter Huber, "Technology: the Great Deregulator," *Forbes,* vol. 15, no. 6, (1995).

19. Desmond S. King, *The New Right: Politics, Markets and Citizenship,* p. 155.
20. Ibid., p. 139.
21. *Corporate Income Taxes in the 1990s,* Institute on Taxation and Economic Policy, October 2000.
22. John Burton, "Taxation Policy and the New Right," in Grant Jordan and Nigel Ashford, eds., *Public Policy and the Impact of the New Right* (London: Pinter, 1993), p. 103.
23. Riddell, Peter, *The Thatcher Era and Its Legacy,* pp. 112–13.
24. Chris Pierson, "Social Policy," in Marquand and Seldon, *The Ideas That Shaped Post-War Britain,* p. 140.
25. Desmond S. King, *The New Right: Politics, Markets and Citizenship,* p. 139.
26. It was former prime minister Harold Macmillan who described it as a sell-off of the family silver: "First of all the Georgian silver goes, and then all the nice furniture that used to be in the salon. Then the Canalettos go." Harold Macmillan, first Earl of Stockton, speech to the Tory Reform Group, November 8, 1985, quoted in Peter Riddell, *The Thatcher Era and Its Legacy,* pp. 27–29.
27. I am using £67 billion to refer to £67,104 million.
28. David Butler and Gareth Butler, *Twentieth Century British Political Facts* (London: Macmillan, 2000), pp. 430–33.
29. John Gray, *False Dawn: The Delusions of Global Capitalism* (London: Granta, 1998), pp. 27–28.
30. Peter Riddell, *The Thatcher Era and Its Legacy,* p. 92.
31. Madsen Pirie, "Reasons for Privatization," in Philip Morgan, ed., *Privatization and the Welfare State,* (1995), p. 26.
32. Peter Riddell, *The Thatcher Era and Its Legacy,* pp. 27–29.
33. Friedrich Hayek, *Law, Legislation and Liberty* (Chicago: University of Chicago Press, 1979), p. 139.
34. David A. Stockman, *The Triumph of Politics* (New York: Harper and Row, 1986), p. 6.
35. Michael Moran and Tony Prosser, "Introduction: Politics, Privatization and Constitutions," in Michael Moran and Tony Prosser, eds., *Privatization and Regulatory Change in Europe* (1994), p. 1.
36. John Gray, *False Dawn: The Delusions of Global Capitalism,* p. 39.

37. Francis Fukuyama, *The End of History and the Last Man* (New York: Routledge, 1992), p. 42.
38. See for example, J. Toye, *Dilemmas of Development: Reflections on the Counterrevolution in Development Theory and Policy* (Oxford: Basil Blackwell, 1993).
39. Vincent Wright, "Industrial Privatization in Western Europe: Pressures, Problems and Paradoxes," in Vincent Wright, ed., *Privatization in Western Europe* (London: Pinter, 1994), p. 3.
40. "Twilight of a God," *The Economist,* September 17, 1994.
41. Edward Carr, "Survey of Business in Europe: What the Ministry Managed," *The Economist,* November 23, 1996.
42. "The Future Surveyed: The Future of Capitalism," *The Economist,* September 11, 1993.
43. "The Post-Soviet World: The Resumption of History," *The Economist,* December 26, 1992.
44. Francis Fukuyama, *The End of History and the Last Man,* p. 41.
45. For a comprehensive review of dependency theory literature, see Robert A. Packenham, *The Dependency Movement: Scholarship and Politics in Development Studies* (Cambridge, MA: Harvard University Press, 1992).
46. *The Times,* December 17, 2001.
47. Robert Skidelsky, "Bring Back Keynes," *Prospect,* May 1997, p. 30.
48. Tony Blair, leader of the opposition, speech to the News Corporation Leadership Conference, Hayman Island, Australia, July 17, 1995.
49. Martin Walker, "No Argument," *Prospect,* March 2000, p. 35.
50. John Gray, *False Dawn: The Delusions of Global Capitalism.*
51. Martin Rhodes, "The Welfare State," in Martin Rhodes, Paul Heyward and Vincent Wright, eds., *Developments in West European Politics* (London: Macmillan, 1997).
52. "Europe's New Left: Free to Bloom," *New Economist,* February 12, 2000.
53. Ibid.
54. Ibid.
55. John Gray, *False Dawn: The Delusions of Global Capitalism,* p. 29.
56. "Europe Wheels to the Right," *The Economist,* May 10, 1997.
57. John Gray, *False Dawn: The Delusions of Global Capitalism,* p. 87.
58. "The Left's New Start: A Future for Socialism," *The Economist,* June 11, 1994.

59. "Displaced, Defeated and Not Sure What to Do Next: The Plight of Europe's Right" *The Economist,* January, 23, 1999.
60. Robert Taylor, "The Social Democrats Come Roaring Back," *New Statesman,* December 20, 1999, p. 26.
61. "Europe's New Left: Free to Bloom," *The Economist,* February 12, 2000.
62. Madsen Pirie, "Reasons for Privatization," in Philip Morgan, ed., *Privatization and the Welfare State,* p. 26.
63. Michael Maclay, "A Mission for Britain," *Prospect,* March 2000, p. 24.
64. Margaret Thatcher, chief opposition spokesman on power, Conservative Political Centre address, Blackpool, October 10, 1968, quoted in Peter Riddell, *The Thatcher Era and Its Legacy,* p. 1.
65. Kenichi Ohmae, *The Borderless World: Power and Strategy in the Global Marketplace* (New York: Harper Business, 1990).
66. "One World?: The Growing Integration of National Economies," *The Economist,* October 18, 1997.
67. Astonishingly, more than 50 percent of Britain's net investment went abroad in 1885–'94, from Kevin O'Rourke and Jeffrey Williamson, *Globalization and History* (Cambridge, MA: MIT Press, 1999).
68. This is redolent of the Barings crisis of 1890 when the merchant bank had to be rescued by the Bank of England after dabbling in South American securities. The same Barings, that is, that went under in 1995 after their Singapore-based trader Nick Leeson bet the wrong way on Japanese equity futures.
69. Globalization theorists like R. Barnet and J. Cavanaugh, *Global Dreams: Imperial Corporations and the New World Order* (New York: Simon & Schuster, 1995), and Raymond Vernon, *Sovereignty at Bay: The Multinational Spread of U.S. Enterprises* (New York: Basic Books, 1971), see the current pattern of MNC expansion to be qualitatively new, whereas globalization sceptics like Hirst and Thompson, *Globalisation in Question: The International Economy and the Possibilities of Governance* (Cambridge, UK: Polity Press, 1996) take a contrary view and claim that an internationalized economy is not unprecedented, that multinationals are not new, and that markets are not necessarily more open or more expansive than they have been through history.
70. William Greider, "Pro Patria, Pro Mundus," *The Nation,* November 12, 2001.

71. World Trade Organization, "Trade and Foreign Direct Investment," 1996; http://www.wto.org/english/news_e/pres96_e/pr057_e.htm.
72. UNCTAD World Development Report 2002.
73. Francis Fukuyama, *The End of History and the Last Man,* p. xiii.
74. Bill Clinton, "The Year 2000 State of the Union: The Economy, Education and Health Care," address by Bill Clinton, President of the United States, delivered to Congress and the nation, Washington, D.C., January 27, 2000.
75. Focus on the Corporation, "Michael Eisner vs. Vietnamese Laborers," March 24, 1998; www.essential.org/monitor.focus/focus.9812.html.
76. Peter Riddell, *The Thatcher Era and Its Legacy,* p. 234.
77. Charlie Leadbeater, "Thatcherism and Progress," in Stuart Hall and Martin Jacques, eds., *New Times: The Changing Face of Politics in the 1990s* (New York: Vers. Books, 1989), p. 396.
78. Office of National Statistics, *Social Trends,* (London, 1999).
79. "Kiwis Turn Sour," *The Economist,* October 19, 1996.
80. "Latin America and the Market: The Free Society on Trial," *The Economist,* November 21, 1998.
81. Paul Krugman, "Some Don't Want to Be Saved From Globalization," *International Herald Tribune,* February 17, 2000.
82. Francis Fukuyama, *The End of History and the Last Man,* p. 41.
83. Debora Spar, "Foreign Investment and Human Rights," *Challenge,* January/February 1999. See also Debora Spar, "China(B): Polaroid of Shanghai Ltd.," Harvard Business School Case Study no. 794–089.

Chapter 3

1. See Giovanni Andrea Cornia, "Inequality and Poverty Trends in the Era of Liberalization and Globalization" (United Nations University Publication, 1999).
2. Naomi Klein, "Cut the Blah Blah Blah," *Guardian,* April 19, 2001.
3. Lamia Kamal-Chaoui, "Halving Poverty," OECD Observer Paris, April 2000; also see UN Human Development Report 2000.
4. For details of South African transition, see W. Munro, V. Padayachee, F. Lund, and I. Valodia, "The State in a Changing World: Plus Ça Change?" *Journal of International Development,* vol. 11, 1999.

5. Lamia Kamal-Chaoui, "Having Poverty."
6. Peter Nolan, *China and the Global Business Revolution* (New York: St. Martin's Press, 2001) and *China and the Global Economy* (New York: St. Martin's Press, 2001).
7. See the UN Human Development Program's 1999 and 2000 Human Development Reports.
8. K. Newland, "Workers of the World, Now What? On the Disappearance of Organized Labor." *Foreign Policy,* Spring 1999.
9. For a review of this literature, see Arik Levinson, "Environmental Regulations and Manufacturers' Location Choices: Evidence from the Census of Manufacturers," *Journal of Public Economics* 62, (1996), pp. 5–29.
10. Support for the "race to the bottom" hypothesis can be found in: T. & F. Collingsworth, William Goold, and Pharis F. Harvey, "Time for a Global New Deal" *Foreign Affairs,* January/February 1994; J. G. Ruggie (1995), "At Home Abroad, Abroad at Home: International Liberalisation and Domestic Stability in the New World Economy," *Millennium,* vol. 24, no. 3, pp. 507–527; Susan Strange, *The Retreat of the State: The Diffusion of Power in the World Economy* (Cambridge: Cambridge University Press, 1996); P. Cerny, "Globalization and the Changing Logic of Collective Action," *International Organization,* vol. 49, (1995), pp. 595–625. However, there is also a body of literature which contradicts this view and even points to the fact that some firms choose to locate in areas of high unionization: J. Friedman, D. Gerlowski, and J. Silberman, "What Attracts Foreign Multinational Corporations? Evidence from Branch Plant Location in the United States," *Journal of Regional Science,* vol. 32, 1992, p. 4; C. Coughlin, J. V. Terza, and V. Arromdee, "State Characteristics and the Location of Foreign Direct Investment Within the United States," *Review of Economics and Statistics,* vol. 73, 1991, pp. 675–83; Timothy J. Bartik, "Business Location Decisions in the United States: Estimation of the Effects of Unionization, Taxes, and Other Characteristics of States," *Journal of Business and Economic Statistics,* vol. 3, no. 1, pp. 14–22. Others suggest that regulatory differences in free trade regimes such as the NAFTA countries only have a minimal impact on firm investment decisions: Gene M. Grossman, and Alan B. Krueger, "Environmental Impacts of a North American Free Trade Agreement," unpublished manuscript prepared for the Conference on the U.S.–Mexico Free Trade Agreement, 1991. Some

also have found no correlation between the intensity of environmental regulation, for example, and the investment decisions of firms; see G. Knogden, "Environment and Industrial Siting," *Zeitschrift für Umweltpolitik.* But as Spar and Yoffie point out (D. L. Spar and D. B. Yoffe, "A Race to the Bottom or Governance from the Top?," in Aseem Prakash and Jeffrey A. Hart, eds. *Coping with Globalization,* (London: Routledge, 2000), pp. 31–51) it is difficult to disaggregate the data and establish causal links and patterns of decision making, therefore it is unclear whether or not corporate movements were or were not spurred by differentials in environmental regulation or wage rates. As they point out, "even if the empirical evidence is somewhat dismissive . . . theoretically they are also quite plausible. Firms undeniably seek to increase profits and create competitive advantage and if moving to less expensive or less onerous locations would serve these aims, it is only logical to expect them to do so."

11. See Robert Wade, "Winners and Losers," *The Economist,* April 28, 2001.
12. "Human Rights Alert," *Business Ethics,* May/June 1999.
13. Centre on Budget and Policy Initiatives and the Economic Policy Institute (2000) in G. Koretz, "Not Enough Is Trickling Down," *BusinessWeek,* January 31, 2000. Within-state income inequality continued to grow in most states in the 1990s despite economic growth and tight labor markets. See also, United for a Fair Economy and Institute for Policy Studies report, "A Decade of Executive Excess: the 1990s: Sixth Annual Executive Compensation Survey."
14. Allan Kennedy, *The End of Shareholder Value* (New York: Orion, 2000).
15. "Pay: Winners and Losers," *The Economist,* May 8, 1999.
16. *Business Week,* November 19, 2001, p. 10.
17. "Divines Opine," *The Economist,* April 12, 1997.
18. P. Barclay, chair., *Income and Wealth, Volume 1: Report of the Inquiry Group* (1995), Joseph Rowntree Foundation report in C. Pantazis and D. Gordon, *Tackling Inequalities: Where Are We Now and What Can Be Done?* (Bristol: The Policy Press, 2000).
19. NBER Research Paper, 7557, *Employment at Will Doctrine.*
20. United Nations Human Development Report (1999), p. 37; http://www.undp .org/hdro/Chapter1.pdf.
21. Complementary massages and free snacks are just some of the perks now on offer at Internet companies.

22. See for example, "Benefits Dwindle for the Unskilled Along with Wages," *The New York Times,* June 14, 1998.
23. The 1999 Human Development Report, UN Development Programme.
24. William Greider, *One World, Ready or Not.*
25. Richard Sennett, *The Corrosion of Character* (New York: Norton, 1998), p. 50.
26. Quoted in H. Martin and H. Schumann, *The Global Trap: Globalization and the Assault on Democracy and Prosperity* (London: Zed Books, 1997); from the *Financial Times,* May 14, 1996.
27. Conversations with Hepatitis C sufferers.
28. IMS Health Report, "Pharmacast and Beyond: A Study of the G7 Antidepressant Market" (1998).
29. Robert Frank, in Juliet Schor, *Do Americans Shop Too Much?* (Boston: Beacon Press, 2000).
30. OECD Employment Outlook, July 1999.
31. International Labor Organization data.
32. Conversations with Internet entrepreneurs.
33. Robert Whymant, *The Times,* February 27, 1998 (Tokyo).
34. R. G. Wilkinson, *Unhealthy Societies: The Afflictions of Inequality* (London: Routledge, 1997).
35. R. G. Wilkinson, "Income, Inequality, and Social Cohesion," *American Journal of Public Health,* September 1997; and McDonough, et al, "Income Dynamics and Adult Mortality in the United States 1972 Through 1989," *American Journal of Public Health,* September 1997.
36. Richard Freeman, "Solving the New Inequality," *Boston Review,* December/January 1997/98.
37. Avery F. Gordon "Globalism and the Prison Industrial Complex: An Interview with Angela Davis," *Race and Class: The Threat of Globalism,* October 1998–March 1999.
38. While it is true that white Americans have been experiencing less crime since 1980, for urban minority Americans who are disproportionately poor, all crimes including homicide are up. The murder rate for inner-city African-American male teenagers and young adults from low-income urban families is about ten times the national average, with minority Americans not only more likely to be the victims of crime but more likely

to be incarcerated as perpetrators. Seven out of every ten inmates locked up in U.S. prisons between 1985 and 1997 were black or Latino.

39. Edward J. Blakely and Mary Gail Snyder, *Fortress America: Gated Communities in the United States* (Washington, D.C.: Brookings Institution Press, 1997).
40. Quoted in Greider, *One World Ready or Not.*
41. Rt. Hon. Stephen Byers's Mansion House speech.
42. Quoted in Alain Lipietz, *La société en sablier,* (Paris: 1996), p. 313.
43. Ulrich Beck, "Beyond the Nation State," *New Statesman,* December 6, 1999, p. 25.
44. Freeman, "Solving the New Inequality."
45. See "The Economist Survey: Globalization and Tax: The Vanishing Taxpayer," *The Economist,* January 29, 2000.
46. Report to the EU: Helsinki Summit, December 1996.
47. Quoted in Ulrich Beck, "Beyond the Nation State."
48. Quoted in Robert Boyes, "Bonn Warned of Cash Flight by Industry," *The Times,* March 2, 1999.
49. Quoted in *The Guardian,* "Left Is Where the Heart Is, Warns Lafontaine," March 15, 1999.
50. "Schröder to Review Tax," *Independent,* March 4, 1999.
51. *The Economist,* August 5, 2000.
52. Ibid.
53. "The Tap Runs Dry: Disappearing Taxes," *The Economist,* May 31, 1997.
54. "Rupert Laid Bare," *The Economist,* March 20, 1999.
55. Krugman, *Fuzzy Math.*
56. Moore and Stansel, *How Corporate Welfare Won: Clinton and Congress Retreat from Cutting Business Subsidies,* CATO Institute Report, 1996.
57. Donald L. Bartlett and James B. Steele, "Paying a Price for Polluters," *Time,* November 23, 1998.
58. Donald L. Bartlett and James B. Steele, "States at War," *Time,* November 9, 1998.
59. Robert Reich, *The Future of Success,* (Portsmouth, NH: Heinemann, 2000), p. 206.
60. "A User's Guide: Investment Agencies," *Corporate Location,* November/December 1996.
61. J. Lloyd-Smith, "Dobson Poised to Approve Use of Relenza in Limited

Cases," *Independent,* October 7, 1999. Note that although the drug wasn't approved, Glaxo did not pull out.

62. Robert Reich, *The Future of Success,* p. 206.
63. *New Republic,* October 29, 2001.

Chapter 4

1. "European Union Report on the NSA," *Daily Telegraph,* December 16, 1997.
2. Adam Smith, *The Wealth of Nations,* 1776.
3. Susan Strange, *The Retreat of the State: The Diffusion of Power in the World Economy.*
4. William Greider, *One World, Ready or Not,* p. 24.
5. Madeline Albright, "First-class diplomacy," statement at confirmation hearing before the Senate Foreign Relations Committee, January 8, 1998.
6. B. Gokay, ed., *The Politics of Caspian Oil* (New York: St. Martin's Press, 2001).
7. The Russian argument was that the Caspian is a closed water system with a fragile ecological balance and all decisions concerning its economic exploitation have to be taken jointly by all states that participate in the Caspian littoral. Russia has demonstrated a "proprietary attitude" toward the Caspian oil deposits and toward energy ventures in the Commonwealth. See "Moscow Pressures Neighbors to Share Oil, Gas Revenues," *Washington Post,* March 18, 1994, p. A24. We should also note that shortly after the dissolution of the USSR, a first interrepublican agreement on distribution of Caspian resources took place with Azeri, Kazakh, Turkmen, and Russian officials at the Ministry of Fuel and Energy in Moscow. See also Angela Spatharou, "Geopolitics of Caspian Oil: The Role of the Integration of the Caspian Region into World Economy in Maintaining Stability in the Caucasus," in B. Gokay, ed., *The Politics of Caspian Oil* (New York: St. Martin's Press, 2001).
8. *Interfax,* September 16, 20, 1994. (Source: *Russian Press Digest* database search).
9. Baku, Azerbaijan Radio, news conference by H. Aliev, September 20, 1994. (Source: *Russian Press Digest* database search).

10. *Interfax,* September 20, 1994. (Source: *Russian Press Digest* database search).
11. Series of articles in the *Financial Times* in 1997: July 18, July 29, August 30, and October 4 when the contracts were actually approved.
12. *Financial Times,* August 30, 1997.
13. *Financial Times,* July 18, 1997.
14. "Britain: Addicted to the Arms Trade," *The Economist,* September 18, 1999.
15. *Financial Times,* August 30, 1997.
16. The USA is the world's biggest arms exporter, accounting for nearly 50 percent of the $53.4 billion (£37 billion) annual market in 1999. (Richard Norton-Taylor, "US Sells Half the World's Arms Exports," *Guardian,* October 20, 2000.)
17. Cited in "Cook's Aide Accuses DTI Over Indonesian Arms," *Independent,* September 17, 1999. See also "Cook Bans Arms for Indonesia," *The Times,* September 26, 1997, and "Britain Still Selling Indonesia Arms," *Independent,* May 15, 1998.
18. http://www.gn.apc.org/tapol/home.htm.
19. Michael Ignatieff, "Human Rights: The Midlife Crisis," *The New York Review of Books,* May 20, 1999.
20. Ahmed Rashid, *Taliban: Militant Islam, Oil and Fundamentalism in Central Asia* (New Haven, CT: Yale University Press, 2001). Also see *The Economist,* December 16, 1995; December 2, 1995.
21. http://www.amnesty.org.uk/action/camp/saudi/repression.shtml.
22. *The Economist,* April 3, 1999.
23. *The Economist,* April 3, 1999; *Financial Times,* April 8, 1999.
24. W. Meyer, "Human Rights and MNCs: Theory vs. Quantitative Analysis," *Human Rights Quarterly,* vol. 18, no. 2. (1996); Han Park, "Correlates of Human Rights: Global Tendencies," *Human Rights Quarterly,* vol. 9 (1987).
25. Debora Spar, "The Spotlight and the Bottom Line," *Foreign Affairs,* March/April 1998.
26. OECD, *Trade, Employment, and Labour Standards* (1996), www.oecd.org.
27. I take the position that there are certain universal human rights that ought to be defended at all costs and that in the defense of them the principle of state sovereignty can and should be overridden.

28. Quoted in Christopher Avery, *Business and Human Rights in a Time of Change* (London: Amnesty International British Section, 2000).
29. Ibid.
30. Derek Brown, "Iraq Sanctions," *Guardian,* August 1, 2000.
31. "China's Snake Charmer—Zhu Returns to China," *The Economist,* April 24, 1999.
32. See for instance *The Times,* October 21, 1999; *Daily Telegraph,* October 25, 1999. Also Gill Bates, "Limited Engagement," *Foreign Affairs,* July/August 1999, vol. 78, no. 4, pp. 65–76.
33. James Baker, "Personal View," *Financial Times,* April 7, 1999.
34. David Acheson, "Personal View," *Financial Times,* June 30, 1999.
35. Gallup poll, June 3, 1999: "Americans' unfavourable attitude towards China unchanged ten years after Tienanmen," www.gallup.com/polls/releases/pr990603.asp.
36. This observation is based upon the time I spent working in Russia in 1991–'93 for the International Finance Corporation and CS First Boston.
37. W. LaFeber, "The Tension Between Democracy and Capitalism During the American Century," *Diplomatic History,* vol. 23, no. 2 (Malden, MA: American Historical Association, 1999).
38. Ibid.
39. T. E. Vadney, *The World Since 1945* (New York: Viking Penguin, 1992).
40. LaFeber, "The Tension Between Democracy and Capitalism During the American Century."
41. Ibid.
42. Ibid.
43. Ibid.
44. Straight to Knox, February 12, 1911, Willard Straight Papers, Cornell University, Ithaca, New York.
45. R. Mokhiber and R. Weissman, "Sanctioning Burma, Sanctioning the United States," Focus on the Corporation, December 23, 1997; http://www.essential .org/monitor/focusfocus.9707.html
46. "Feeding Mammon—World Trade Organisation: Special Report," *Guardian,* November 30, 1999.
47. Mark Lynas, *The World Trade Organisation and GMOs,* Consumer Policy Review, London, November/December 1999.

48. "Feeding Mammon—World Trade Organisation: Special Report," *Guardian,* November 30, 1999.
49. "Media Advisory: Initial Reports from Seattle Gloss Over WTO issues," *Fairness and Accuracy in Reporting* (FAIR), December 1, 1999. Also see Carol Miller and Jennifer L. Croston, "WTO Scrutiny v. Environmental Objectives: Assessment of the International Dolphin Conservation Program Act," *American Business Law Journal,* vol. 37 (1999). See also "Focus: Trade Wars, the Hidden Tentacles of the World's Most Secret Body," *Independent on Sunday,* July 11, 1999. See also Mark Lynas, "The World Trade Organisation and GMOS," *Consumer Policy Review,* November/December 1999.
50. David Korten, *When Corporations Rule the World* (San Francisco, CA: Berret-Koehler, 1995).
51. Carol J. Miller and Jennifer L. Croston, "WTO Scrutiny v. Environmental Objectives: Assessment of the International Dolphin Conservation Program Act," *American Business Law Journal.*
52. "Focus: Trade Wars, the Hidden Tentacles of the World's Most Secret Body," *Independent on Sunday;* Clarke, "The Superpower Elite that Sidelines the Poorest Nations," *Independent on Sunday.*
53. John Madeley, "There's a Food Fight in Seattle," *New Statesman,* November 22, 1999. It was finally resolved, however, that Supachia would share the post with Michael Moore. Supachia was in line to take over from Moore in 2002 for a further three years.
54. Mark Lynas. And see also George Monbiot, "Britain and America Have Given Big Business an Inhuman Bonus," *Guardian,* October 28, 1999.
55. Mark D. Fefer, "Not-So-Free Trade," *Seattle Weekly,* April 22–28, 1998. Note that the organization that solicited large contributions from the private sector for preparations for the Seattle WTO round, the Seattle Host Organization, was cochaired by Bill Gates and Boeing CEO Phil Condit.
56. Mark Lynas.
57. See also Marceau Gabriell and Peter N. Pederson, "Is the WTO Open and Transparent?," *Journal of World Trade,* February 1999.
58. Mark Milner and Stephen Bates, "Trade Rules Flouted in Banana Dispute," *Guardian,* March 11, 1999.
59. Donald E. Barlett and James B. Steele, "How the Little Guy Gets Crunched," *Time,* February 7, 2000.
60. Ibid.

61. World Development Movement "DBCP Legal Action Report," August 1997.
62. William Greider, "Pro Patria, Pro Mundus," *The Nation,* November 12, 2001. From D. Cohn (2000) "Globalization and the Possibility of Human Agency," Canadian Public Administration 43(4) 490502.
63. Samuel Hungtingdon, "The Erosion of American National Interests," *Foreign Affairs,* September/October 1997.

Chapter 5

1. Michael Ellison and Martin Kettle, "Back-door Donations Shatter All Spending Records," *Guardian,* November 7, 2000.
2. "Money and Politics: Politicians for Rent," *The Economist,* February 8, 1997.
3. Ibid.
4. Margaret Scammell, "Political Marketing: Lessons for Political Science," *Political Studies,* XLVII, p. 720.
5. Peter Marks, "The 2000 Campaign: The Ad Campaign," *New York Times,* June 13, 2000.
6. Martin Harrop, "Political Marketing" in *Parliamentary Affairs,* vol. 43, no. 3. Quoted in Margaret Scammell, *Designer Politics: How Elections Are Won* (Palgrave, 1995), p. 3.
7. Margaret Scammell, *Designer Politics,* p. 4.
8. Charles Lewis, "Keeping Government Accountable," Centre for Public Integrity, speech to the Institute of Security Studies, April 12, 1999.
9. Julian Borger, "For Sale: The Race for the White House," *Guardian,* January 7, 2000.
10. Ibid.
11. "Fuelling the Political Machine," *The Economist,* November 15, 1997.
12. Paul Heywood, "Political Corruption: Problems and Perspectives," *Political Studies,* vol. 45, no. 3 (1997), pp. 430–31.
13. Charles Lewis, "Capital Gains on Capitol Hill," Centre for Public Integrity (2000); http://www.publicintegrity.org/PressRoom_Fatcat.html.
14. See "Are Corporations/Businesspeople Abusing the Power They Wield in Politics?," *Forbes,* February 8, 1999; "Do Corporate PAC's Restrict Competition?," *Business and Society Journal,* Chicago, June 1998.

15. Donald E. Barlett and James B. Steel, "How the Little Guy Gets Crunched," *Time,* February 7, 2000.
16. A report from the U.S. General Accounting Office in 1996 put the cost to the taxpayer at $126.4 billion for resolution costs and $285.2 billion for interest expenses, a total of $411.6 billion.
17. "You Pays Your Money," *The Economist,* July 31, 1999.
18. Centre for Public Integrity, "The Buying of the President," (2000); http://www.publicintegrity.org/buying_question.html. See also John Bores, "Money Business and the State: Material Interests, Fortune 500 Corporations and the Size of the Political Action Committee," *American Sociological Review,* vol. 154, pp. 821–33.
19. Lars-Erik Nelson, "Democracy for Sale," *New York Review of Books,* December 3, 1998.
20. Julian Borger, "For Sale: The Race for the White House."
21. Ibid.
22. Lars-Erik Nelson, "Democracy for Sale," *New York Review of Books,* December 3, 1998.
23. Jeffrey Birnbaum, "Microsoft's Capital Offense," *Fortune,* February 2, 1998.
24. Center for Responsive Politics, September 6, 2001: http://www.opensecrets.org/alerts/v6/alertv6_26.asp.
25. Julian Borger, "Power Firm Vetted Bush's Energy Regulators," *Guardian,* May 26, 2001.
26. Julian Borger, "All the President's Businessmen," *Guardian,* April 27, 2001.
27. "U.S. and Canada Satellite Companies Accused of Inadequate Safeguards," *Financial Times,* May 26, 1993, p. 3. See also "Unlovey-Dovey: China's Blocked Satellite Deal," *The Economist,* February 27, 1999. Paul Mann, "High-tech Dispute Dogs U.S./Sino Summit," *Aviation Week Space Technology,* May 11, 1998.
28. Jean Charles Brisard and Guillaume Dasquie, *Bin Laden—The Forbidden Truth.*
29. Lars-Erik Nelson, "Democracy for Sale," *New York Review of Books,* December 3, 1998.
30. Andrew Jack, "Struggle to Get Pact Banning Bribery," *Financial Times,* November 21, 1997.

31. *Guardian,* December 4, 1999.
32. *The Economist,* April 10, 1999.
33. Hans-Dieter Klingemann, "Mapping Political Support in the 1990s: A Global Analysis," in Pippa Norris, ed., *Critical Citizens* (1999), p. 50.
34. "Money and Politics: Politicians for Rent," *The Economist,* February 8, 1997.
35. Ibid.
36. John Curtice and Roger Jowell, "The Sceptical Electorate," *British Social Attitudes, the 12th Report,* Social and Community Planning Research (1995), p. 141.
37. Transcript from the legal case of *Mostyn Neil Hamilton* v. *Mohamed Al Fayed,* November 19, 1999.
38. Lars-Erik Nelson, "Democracy for Sale," *New York Review of Books,* December 3, 1998.
39. "The Globalisation of Communications," *The Economist,* November 29, 1997.
40. Nicholas Watt, "Soccer Decision Leaves Murdoch's Backing for New Labour in Doubt," *Guardian,* April 10, 1999.
41. *Sun,* March 18, 1997.
42. "The Sun Also Sets," *The Economist,* March 22, 1997.
43. Quoted in Richard Belfield, Christopher Hird, and Sharon Kelly, *Murdoch: The Decline of the Empire* (London, 1991).
44. Ibid.
45. S. Kaplan, "Payments to the Powerful," *Columbia Journalism Review,* September/October 1998.
46. Greg Miller, "Hollywood Sees a Pro-business Ally in Bush," *Los Angeles Times,* January 19, 2001.
47. Quoted in David Korten, *When Corporations Rule the World.*
48. Matthew Josephson, *The Robber Barons: The Great American Capitalists, 1861–1901* (New York: Harcourt Brace and Company, 1934).
49. Russell, J. Dalton, "Political Support in Advanced Industrial Democracies," in Pippa Norris, ed., *Critical Citizens* (Oxford: Oxford University Press, 1999), p. 74.
50. See *Gallup Political and Economic Index,* 1991–'99.
51. In the 1998–'99 Gallup surveys, health, unemployment, and education scored consistently high as the most urgent problems facing Britain. In a

survey of bills debated in the House of Commons in the 1998–'99 session, only twenty-two out of a total of 144 (15 percent) related to these issues.

52. The Gallup Organization—Poll Releases, January 3, 1997; www.gallup.com/poll/releases/pr970103.asp.
53. Robert D. Puttman, *Bowling Alone,* (London: Simon & Schuster, 2000), p. 47.
54. Henley Centre, *Planning for Social Change, 1996–97* (1996), quoted in Mark Leonard, *Britain*™ (1997), p. 28.
55. World Values Surveys, quoted in Pippa Norris, "Institutions and Political Support," in Pippa Norris, ed., *Critical Citizens,* p. 229.
56. Susan J. Pharr, and Robert D. Putnam, *Disaffected Democracies: What Is Troubling the Trilateral Countries?* (Princeton: Princeton University Press, 2000).
57. *Voter Turnout from 1945 to 1997: A Global Report on Political Participation,* pp. 81 and 94.
58. The Pew Research Çenter for the People and the Press, "Few See Choice of President as Important," June 2000.
59. Charles Lewis, "Keeping Government Accountable," Centre for Public Integrity, speech to the Institute of Security Studies, April 12, 1999.
60. *Financial Times,* June 15, 1999.
61. Susan E. Scarrow, *Parties and Their Members: Organizing for Victory in Britain and Germany* (Oxford: Oxford University Press, 1996), p. 57.
62. *L'Etat de la France 1993/94,* (Paris, 1993), quoted in Jeremy Richardson, "The Market for Political Activism: Interest Groups as a Challenge to Political Parties," *West European Politics,* vol. 18, no. 1, p. 121.
63. Susan E. Scarrow, Center for German and European Studies Working Paper 2, 59, University of California, Berkeley (1996).

Chapter 6

1. "Protests Against 'Corporate Evil' Take Root Across Small-town America," *Financial Times,* August 8, 2000.
2. Helmut Anheier, Marlies Glasius, and Mary Kaldor, eds., *Global Civil Society* (Oxford: Oxford University Press, 2001).
3. Peter Melchett, *New Statesman,* January 10, 2000, p. xiii.

4. Tony Jackson, "The Global Company: Facing Up to a Challenging Opposition," *Financial Times,* October 31, 1997, p. 16.
5. Helmut Anheier, Marlies Glasius, and Mary Kaldor, eds., *Global Civil Society.*
6. Christopher L. Avery, *Business and Human Rights in a Time of Change* (London: Amnesty International British Section, 2000).
7. *New Start Worship,* New Start, 2000.
8. www.thehungersite.com.
9. Nancy Dunne and Stella Burch, "Clinton Moves on Sweatshops," *Financial Times,* August 5, 1996, p. 3.
10. Neil Buckley, "Rise of the Ethical Consumer," *Financial Times,* April 27, 1995.
11. Co-op America, Harwood Group.
12. Roger Cowe, "Caring Attitudes Out of Fashion," *Guardian,* March 12, 1999.
13. Tony Jackson, "The Global Company: Facing Up to a Challenging Opposition," *Financial Times,* October 31, 1997, p. 16.
14. "The Enron Corporation: Corporate Complicity in Human Rights Violations," Human Rights Watch, January 1999. Although this was not the first time that Human Rights Watch had condemned the role of corporations, the Enron report was distinct in that the corporation was the focus of the report and named in its title.
15. Peter Melchett, *New Statesman,* January 10, 2000.
16. John Vidal, "Consumer Power Ready to Take on Corporations," *Guardian,* November 27, 1999; "Germany: Market for Green/Ethical Investment Products Booms," Handelsblatt.
17. Thomas Bauer, "New Market Report: Green's Shares and Funds—A Growth Area," *Öko-Zentrum NRW,* January 23, 1999.
18. Richard Johnson and Daniel Greening, "The Effects of Corporate Governance and Institutional Ownership Types on Corporate Social Performance," *Academy of Management Journal,* October 1999.
19. "Pru to Put Investments Through Ethical Screening," *The Times,* November 24, 1999.
20. Conversations with CEOs, 2000–2001.
21. Debora Spar, "Foreign Investment and the Pursuit of Human Rights," *Challenge,* January/February 1999.

22. Conversations with the author.
23. Debora Spar, "The Spotlight and the Bottom Line," *Foreign Affairs,* March/April 1998.
24. A. Maitland, "The Value of Virtue in a Transparent World," *Financial Times,* August 5, 1999.
25. Joanna Bale and Valerie Elliott, "Byers Calls on Consumers to Use their Power," *The Times,* July 23, 1999. See also White Paper, "Modern Markets: Confident Consumers," Stephen Byers, July 1999.
26. Ibid.

Chapter 7

1. Quoted in DFID, *Viewing the World,* July 2000.
2. Robin Anderson, *Consumer Culture and TV Programming* (Boulder, CO: Westview Press, 1995).
3. "News Going Nowhere," *Observer,* August 20, 2000.
4. Anderson, *Consumer Culture and TV Programming.*
5. "Turning Off the Presses," *The Economist,* October 11, 1997.
6. Anderson, *Consumer Culture and TV Programming.*
7. Reuters, "Journalists' Federation Warns on AOL-Time Merger," January 11, 2000.
8. Lynda Morris, *The Ethical Consumer: A New Force in the Food Sector,* Leatherhead Food Research Association, executive summary (1996).
9. "Sainsbury's in Bid to Develop New GM Foods," *Sunday Times,* June 4, 2000.
10. A. Maitland, "The Value of Virtue in a Transparent World," *Financial Times,* August 5, 1999.
11. Institute of Business Ethics, Company Use of Codes of Business Conduct, (1998).
12. Sustain-Ability, *The Oil Sector Report: A Review of Environmental Disclosure* (1999).
13. KPMG, *International Survey of Environmental Reporting* (1999).
14. B. Dennis, C. Neck, and M. Goldsby, "The Body Shop International: An Exploration of Corporate Social Responsibility," *Management Decision,* vol. 36, no. 10 (1998).
15. Z. Monkheiber, "Science for Sale," *Forbes,* May 17, 1999.

16. Emma Brockes and Julian Borger, "Tiger Trap," *Guardian,* July 26, 2001.
17. Borger, Julian, "All the President's Businessmen," *Guardian,* April 27, 2001.
18. Z. Monkheiber, "Science for Sale."
19. See, for example, "Think Tanks Sell Firms EU Access," *Sunday Times,* April 2, 2000.
20. K. Maguire, "University Takes Tobacco 'Blood Money'," *Guardian,* December 7, 2000.
21. "Focus on the Corporation," April 2, 1998; www.essential.org/monitor/focus.
22. "Developing a Global Response—Legislating for a Tobacco-free Society," presentation given by Derek Yach, Executive Director, Noncommunicable Diseases and Mental Heath, World Health Organization, 2001.
23. Ron Nixon, "The Corporate Assault on the Food and Drug Administration," *International Journal of Health Services,* vol. 26, no. 3, (1996), pp. 561–8.
24. Conversation with CEO.
25. See, for example, www.responsibleshopper.org.
26. The Boycott Board and Microsoft Boycott Campaign; www.boycott.2street.com; www.vcnet.com.
27. www.corpwatch.org, www.igc.org, and www.mcspotlight.org.
28. Several states still have suits pending, which may, depending on the outcome, impact upon Microsoft's ability to continue doing so. Rivals such as AOL have launched civil suits, and the European Union is in the process of an antitrust case against Microsoft.
29. Pippa Norris, *Democratic Phoenix: Political Activism Worldwide* (Cambridge: Cambridge University Press, 2002).
30. 6,750 on www.google.com.
31. *Guardian,* August 27, 1998.
32. Diane Summers, "Charities Accused of Exaggerating Advert Claims," *Financial Times,* October 4, 1995, p. 11.
33. Michael Shaw Bond, "The Backlash Against NGOs," *Prospect,* April 2000.
34. Frank Furedi, "It's Just a Failure of Nerve," *New Statesman,* January 10, 2000, p. xxviii.

35. Wyn Grant, *Pressure Groups, Politics and Democracy in Britain,* 2nd ed., (London: 1995).
36. Lynda Morris, *The Ethical Consumer: A New Force in the Food Sector,* Leatherhead Food Research Association, executive summary (1996).
37. National Consumer Council, Annual Review 1999; www.ncc.org/underst.htm
38. Neil Buckley and Peter Marsh, "Tesco Bows to Plastic Pressure," April 25, 1995, p. 12.
39. *A Sporting Chance: Tackling Child Labour in India's Sports Goods Industry,* Christian Aid Report, 1997.
40. *The Apparel Industry and Codes of Conduct: A Solution to the International Child Labour Problem?* Bureau of International Labor Affairs, U.S. Department of Labor, 1996.
41. *Child Labour: The Way Business Can Help Children,* Anti-Slavery International, 1996.
42. Committee of Inquiry, *A New Vision for Business* (1999), p. 54; www.business-impact.org.
43. Frank Furedi, "It's Just a Failure of Nerve," *New Statesman,* January 10, 2000, p. xxviii.
44. Robert Shrimsley, "Customer Power Will Help Tackle 'Rip-off' Britain," *Daily Telegraph,* July 23, 1999.

Chapter 8

1. There are, of course, exceptions, such as Steve Rubin and Silvio Berlusconi.
2. Porter Bibb, *Ted Turner's Amazing Story; It Ain't as Easy as It Looks* (New York: Crown, 1994), p. 415.
3. "Ben & Jerry & Nato," *New York Times,* April 28, 1998.
4. "Military Excess from the Executive Office," June 13, 1999. "NATO Expansion Foes Rally for Senate Votes—Debate in Eastern European Countries," *Washington Post,* April 27, 1998. "Large Scoops of Social Values," *Financial Times,* February 9, 1998.
5. "Clinton's Defense Budget Hike Targeted—Many Consider His Spending Priorities Out of Place," *San Francisco Examiner,* January 30, 1999.

6. T. Emerson, "A Letter to Jiang Zemin," *Newsweek International,* May 29, 2000.
7. Conversation with Paul Spicer, former Deputy Chairman, Lonrho.
8. The U.K. suspended its sanctions against Libya in July 1999 when the Lockerbie suspects were extradited to stand trial. A decision to permanently suspend the sanctions will not be taken until after the verdict.
9. See Noreena Hertz and Michael Porter, "The Bottom Up Solution," *Financial Times,* September 16, 1997.
10. Conor O'Clery, *Greening of the White House* (Dublin: 1996).

Chapter 9

1. "Australian Business Intelligence: HIV Sweeps Through the Mining Industry," *Sydney Morning Herald,* November 26, 1999.
2. Schuettler, "AIDS Threatens to Bury SA Mining Industry," July 28, 1999; http://www.wozainternet.co.za.
3. Bekker, Jacoline, "The Weave of Business and Government" (1998), Judge Institute of Management Studies, unpublished dissertation.
4. The Shell Petroleum Development Company of Nigeria Limited, "1999: People and the Environment Annual Report."
5. "Nigerian Tribes Cut Oil Lines," *The Times,* February 22, 1999.
6. *The New York Times,* July 1999.
7. John Browne and Reith Lecture, *The Times,* April 29, 2000.
8. J. Bennet, "Franchises for Inner City: Economic Development Tool," *Planning,* November 1998.
9. M. Hickins, "Rite Aid in the City," *Management Review,* March 1999.
10. At the last local British elections in 1999, the Benwell ward of Newcastle, one of the U.K.'s most deprived inner-city areas, an area where three quarters of children live in households with no earned income, saw a turnout of only 19.4 percent.
11. More than 98 percent of British schools participate in Tesco's Computers for Schools and Walkers' Free Books scheme.
12. Eagle Rock brochure—American Honda Education Corporation.
13. "The High School at the End of the Road," *New York Times Magazine,* July 5, 1998.
14. "Analysis: Corporate Responsibility: The Cannibals Adopt Cutlery,"

Guardian March 18, 1999; and BT, "Changing Values," http://www.bt.com/world/sus.dev/; 1998.

15. "Beyond the Grey Pinstripes," World Resources Institute and the Aspen Institute, October 2001.
16. See for example Walker Information and Council of Foundations. "Measuring the Business Value of Corporate Philanthropy," http://www.walkerinfo.com or "The Corporate Citizen: Adopting a Socially Responsible Image," *Crossborder Monitor,* September 28, 1994.
17. Chris Marsden and Jörg Andriof, "Towards an Understanding of Corporate Citizenship and How to Influence It," *Journal of Citizenship Studies,* July 1998.
18. Ibid.
19. The Cone/Roper Corporate Citizenship Study, November 2001.
20. 1997 Access Omnibus Survey by Business and the Community.
21. "Millennium Poll on Corporate Society Responsibility," Environics, 1999.
22. "85 Percent of Corporations Use Cause Marketing Tactics," *Direct Marketing,* September 2000.
23. "Oil Firm's Secret Deal to Free Burma Prisoner," *The Times,* December 7, 2000.

Chapter 10

1. Julian Borger, "Coca-Cola Sued Over Bottling Plant 'Terror Campaign,' " *Guardian,* July 21, 2001.
2. Premier Oil, for example, is playing the role of partner and conduit in Burma to NGOs such as Amnesty International and Save the Children, who otherwise could not legally function in the country.
3. "Where There's Smoke," *Guardian,* September 18, 2000.
4. Stephen Glass, "Gift of the Magnate," *New Republic,* January 26, 1998.
5. Ibid.
6. For a good summary of the general problems of depending on philanthropy, "A Major Transfer of Government Responsibility to Voluntary Organizations. Proceed with Caution," *Public Administration Review,* Washington, May/June 1996.

7. Charles Leadbeater and Geoff Mulgan, "Lean Democracy and the Leadership Vacuum," *Lean Democracy,* (1994), p. 14.
8. I. Fuyuno, "A Silent Epidemic," *Far Eastern Economic Review,* Hong Kong, September 28, 2000, vol. 163, no. 39, pp. 78–80.
9. Virgin Atlantic Airways, which had just started flights to the Caribbean, stepped in.
10. Already started: the computers come with the MS Encarta Encyclopedia and other MS education software.
11. Naomi Klein, *No Logo: Taking Aim at the Brand Bullies* (London: Picador, 2000).
12. A few countries recognize the potential conflicts, and regulate marketing to children so as to disallow such practices. Denmark prohibits any kind of commercially sponsored material or activity in schools. In Belgium companies can produce materials, but schools are prohibited from using them. Most countries leave industry to regulate itself.
13. Naomi Klein, *No Logo: Taking Aim at the Brand Bullies.*
14. Ibid.
15. "AOL, *Time,* Cite Social Goals," Yahoo! headlines, January 11, 2000; http://uk.news.yahoo.com/000111/22/dayw.html.
16. In the United States, companies face a significant liability if their shareholders believe that they are not making the best use of their investment by putting it toward social causes, and investors can sue for damages.
17. BP Harvard Business School Case Study: F. Reinhardt, *Global Climate Change and BP Amoco,* (Cambridge, MA: 2000).

Chapter 11

1. John S. Dryzek, "Transnational Democracy," *Journal of Political Philosophy,* vol. 7, no. 1 (1999), p. 38.
2. Thomas L. Friedman, *The Lexus and the Olive Tree* (New York: Farrar, Straus & Giroux, 1999): Thomas Friedman puts forth "the Golden Arches Theory of Conflict Prevention," which states that "no two countries that both had McDonald's had fought a war against each other since each got its McDonald's."
3. Jessica Matthews, *Foreign Affairs.*

4. Sydney Tarrow, "Mad Cows and Social Activists," in Susan J. Pharr and Robert D. Putnam, eds., *Disaffected Democracies,* p. 286.
5. John S. Dryzek, "Transnational Democracy," pp. 45–46.
6. "The Non Government Order," *The Economist,* December 9, 1999.
7. For an illuminating description of the Seattle protest, see Jeffrey St. Clair, "Seattle Diary: It's a Gas, Gas, Gas," *New Left Review,* no. 238, November/December 1999.
8. "Housewives' Revolt Rocks Japan," *The Times,* July 22, 2000.
9. W. Hutton and A. Giddens, eds., *On the Edge: Living with Global Capitalism* (London, 2000).
10. John S. Dryzek, "Transitional Democracy," pp. 45–46.
11. Richard Hofstadter, ed., *The Progressive Movement 1900–1915,* (Englewood Cliffs, NJ: Prentice Hall, 1963), p. 2, and J. A. Thompson, *Progressivism,* British Association for American Studies (1979), p. 7.
12. Ibid., p. 349.
13. Ibid., p. 349.
14. Richard L. McCormick, *The Party Period and Public Policy,* (Oxford: Oxford University Press, 1986), pp. 326–27.
15. Ibid., p. 377.
16. Ibid, p. 327.
17. J. A. Thompson, *Progressivism,* p. 37.
18. Richard McCormick, *The Party Period and Public Policy,* p. 274.
19. Ibid., p. 283.
20. Ibid., p. 276.
21. Ibid., p. 272.
22. Rachel Carson, *Silent Spring,* (Boston: Houghton Mifflin, 1962).
23. Russell Dalton, *The Green Rainbow,* (New Haven, CT: Yale University Press, 1994), p. 90.
24. Ibid.
25. "Greens Grow Up," *The Economist,* August 7, 1999.
26. "I am very sympathetic with a lot of the causes being raised by all the people that are there demonstrating." Bill Clinton, December 1, 1999. Quoted in *The Times,* December 2, 1999.

ACKNOWLEDGMENTS

I would like to thank Professor Robin Matthews and Professor Debora Spar for their insightful comments on earlier drafts and unstinting support for the project, and also Charles Hampden Turner, Professor Peter Nolan, Professor John Child, Professor Sandra Dawson, Christos Pitelis, Jamie Mitchell, Richard Symons, Tanya Schwarz, and Jamie Miller for their intellectual input.

Catherine Needham and Beth Ahlering for their excellent research assistance and enthusiasm for the project and also Angela Spatharou, Chryssa Kanellakis-Reimer, Christopher Garner, Robert Sabbarton, Nadia Kokourina, Becky Morris, Tom Gross, Raj Lalsare, Virginia Graham, Paul Davies, and Rob Koepp for their help on various chapters.

Gillon Aitken, David Lloyd, and Samir Shah for believing in me and the project in its most early days, and Clare Alexander for providing me with support and care all the way through.

All those at Random House U.K., especially my editor Ravi Mirchandani, for helping to make the book the great success it has been at home.

Fred Hills, my editor at Free Press, whose enthusiasm for the project has been much appreciated, and whose suggestions for the

U.S. edition are much valued. Marion Maneker for bringing *The Silent Takeover* to HarperBusiness.

Julia Hobsbawm, Sir Martin Sorrel, and John Lloyd for their encouragement. Len Blavatnik and Henry Porter for their overwhelming generosity and support.

Pat Lilley for keeping me organized, Andrew Flower for keeping me well, and Ralph Hancock for his eagle eyes.

My friends—who put up with the fact that they did not see me for months on end and not only rarely complained but continued to phone. Especially Orson for making the writing period a magical time; Alexis for continuing to be an incredibly important part of my life and always being there when I need him; James for those special treats that did make the bad bits better; and Juliet for her kindness.

INDEX

ABOUT THE AUTHOR

Author, academic, and broadcaster, Dr. Noreena Hertz is the Associate Director of the Centre for International Business at the University of Cambridge. She began her career as a Russia expert, and she worked for the World Bank in 1992 advising the Russian government on its economic reforms. In the mid-nineties she worked on the Middle East Peace Process with the Palestinian Authority and the governments of Israel, Jordan, and Egypt.

With the critically acclaimed publication of *The Silent Takeover* in Europe, Noreena Hertz has become recognized as one of the world's leading young experts on economic globalization. Her op-ed pieces have been published in the *Washington Post,* the *New Statesman,* the *Observer,* and the *Guardian,* and she is a regular commentator on both television and radio.

The Silent Takeover has been translated into French, German, Chinese, Czech, Spanish, Italian, Dutch, Korean, Portuguese, and Japanese. The author received her M.B.A. from Wharton and a B.A. in philosophy and economics from University College London. Her Ph.D. is from the University of Cambridge. She lives in London.